Hypertext

Hypertext

The Electronic Labyrinth

Ilana Snyder

NEW YORK UNIVERSITY PRESS
Washington Square, New York

First published 1996 by Melbourne University Press

Text © Ilana Ariela Snyder 1996, 1997
Design and typography © Melbourne University Press 1996

First published in the U.S.A. in 1997 by
NEW YORK UNIVERSITY PRESS
Washington Square
New York, N.Y. 10003

Printed in Malaysia

CIP data available from the Library of Congress
ISBN 0-8147-8078-4 (clothbound)
ISBN 0-8147-8079-2 (paperbound)

For Dulcie and Abe

Contents

Preface

Hypertext is an information medium that exists only on-line in a computer. A structure composed of blocks of text connected by electronic links, it offers different pathways to users. Hypertext provides a means of arranging information in a non-linear manner with the computer automating the process of connecting one piece of information to another. When the structure accommodates not only printed texts but also digitised sound, graphics, animation, video and virtual reality, it is referred to as 'hypermedia'. 'Multimedia' is often adopted as a marketing term by computer manufacturers, software developers, publishers and others to describe both hypermedia content and the hardware or software that embodies it. Hypertext is my preferred term. I use it to represent both the structure and the content of this type of information technology.

A hypertext is constructed partly by the writers who create the links, and partly by readers who decide which threads to follow. Unlike printed texts, which generally compel readers to read in a linear fashion—from left to right and from top to bottom of the page—hypertexts encourage readers to move from one text-chunk to another, rapidly and non-sequentially. Hypertext differs from printed text by offering readers multiple paths through a body of information: it allows them to make their own connections, to incorporate their own links and to produce their own meanings.

Hypertext consequently blurs the boundaries between readers and writers. These differences help to support the view that the use of hypertext affects how we read and write, how we teach reading and writing, and how we define literacy practices.

I am particularly interested in the close connections between hypertext and certain key ideas in deconstruction, structuralism, post-structuralism, reader-response theory, narratology and critical literacy. Hypertext is also linked with different—possibly enhanced—ways of teaching and learning in the humanities and social sciences. Some argue that it enables students to understand fundamental aspects of contemporary literary theory. Some think that those who explore a hypertext system and make their own connections in an ever-expanding web of information are engaged in the central activities of education: namely, constructing knowledge and developing independent critical thinking. Others claim that, although hypertext is still in its infancy, it may influence intellectual development as profoundly as the invention of alphabetic writing did in the thirteenth century BC or the printed book in the fifteenth century AD. Hypertext and the other electronic reading and writing technologies are seen as having reached that critical mass which will enable them to supersede the printed word as the dominant means of enculturation and communication.

This book examines the substance and viability of such arguments, claims and beliefs. I neither assume much technological expertise and experience nor provide a technical description of the electronic medium. I alert readers, however, to the ways in which technological determinism permeates academic discourse about technology. By 'technological determinism' I mean the assumption that qualities inherent in the computer medium itself are responsible for changes in social and cultural practices. Hypertext is often discussed in a celebratory if not hyperbolic manner. We read that hypertext is replacing linear writing in an evolutionary step towards a perfect communication technology; that the mere act of linking multiple interpretations and voices results automatically in better communication; and that hypertext is transforming society and education systems, democratising the academy and promoting the breakdown of artificial divisions between the disciplines. Such grandiose claims need to be interrogated assiduously, since they build on the premise that technology is directly responsible for changes that necessarily enhance social relations. Overlooking the human agency integral to all technological innovation, they rely on an interpretative frame in which any notion of control over technology disappears.

The functions of hypertext are not wholly determined either by technology or society. As Selfe and Selfe (1994:482) point out, 'computers, like other complex technologies, are articulated in many ways with a range of existing cultural forces and with a variety of projects in our educational system . . . that run the gamut from liberatory to oppressive'. Moreover, seeing that the humanities and the social sciences have been using hypertext technology for only ten years or so, claims about the nature of its impact can only be provisional. While it is salutary to investigate the ways in which hypertext may illuminate literary theory and enhance literacy practices, to hail its advent as the beginning of a social and educational revolution is politically naive.

The quandary in writing about hypertext is how to describe in a linear text an anti-linear style of writing and technology. This replicates the difficulty of encoding one technology within another: it could be argued that to understand hypertext you must experience it. Bolter (1991) and Lanham (1993) have responded to this conundrum by making their texts available in two formats— as both printed book and hypertext. This option, however, was not available to me in 1995.

Terminology is another difficulty. Since hypertext alters the experiences associated with reading, writing and textuality, it is problematic to describe it in terms so closely connected with print technology. Because the roles of both author and reader are reconfigured in the hypertext medium, current terminology is not surprisingly limited. The term 'text', for example, is particularly troublesome. Electronic (or 'virtual') textuality differs from print textuality. Whether converted from print to electronic form or created wholly in an electronic environment, such 'texts' display characteristics quite distinct from those taken for granted in the print medium. The difficulties associated with writing about hypertext are not easily overcome. To highlight them indicates just how inadequate traditional modes of communication are for describing those new modes which are altering our understanding of literacy practices.

Fascinated by hypertext, I remain sceptical of some of its putative possibilities. Avoiding the excesses of techno-evangelism, I aim to describe the technology of hypertext, its connections with contemporary literary theories and its implications for education. My position approximates what Aronowitz (1992:121) calls 'postcritical'. It differs from the characteristic stance of enthusiasts like Papert (1980) and Zuboff (1984), who embrace the computer as a tool that can become a vehicle of human creativity, or critics like Weizenbaum (1976) and Ellul (1980), who warn that computers have penetrated not only our social relations but also our personalities and culture. By contrast, proponents of a post-critical perspective argue that it is futile to deplore the influence of electronic mediations on everyday life, because whether we like it or not we live now in a 'technoculture' (Aronowitz, 1992:121). Our challenge is to negotiate the problems of becoming literate in the computer-mediated culture of the future. My personal interest is in how the versatile and volatile technology of hypertext may be used educationally for imaginative and playful purposes.

Acknowledgements

This book is itself a kind of hypertext created out of the connections I have made between the ideas of key theorists in the area of electronic literacy. It is to those writers that I owe thanks. I also owe thanks to John Iremonger, a self-confessed computer illiterate, for suggesting that I write the book, to Michael Joyce for his generous response to a draft, to Ken Ruthven for his warm encouragement and meticulous editing, to Susan Keogh for her careful checking of the final manuscript, and to the Faculty of Education, Monash University, for providing me with a stimulating environment in which to work and write. I owe special thanks to my colleagues Margaret Gill, Dick Selleck and Colin Evers for their continued interest in my projects. As well, I owe thanks to the Australian Association for Research in Education (AARE) for awarding me a Travelling Research Fellowship in 1994. The research seminars at four universities in New South Wales and Queensland provided me with an invaluable opportunity to rehearse my ideas about hypertext as they evolved. Above all, I am grateful to my husband Ray and two sons, Gabe and Ben, for their love and support throughout my two-year obsession with Hypertext.

1

Electronic Writing

Technologies of writing

It is easy to recognise the printing press and the typewriter as technologies of writing because both involve mechanisation. Yet the papyrus roll and the vellum manuscript also exemplify technologies of writing, both of which required devices: the reed pen and papyrus in ancient Egypt, and the quill and parchment in the Middle Ages. Working with materials such as plants and animal skins to produce a finished volume or book demanded considerable technical knowledge. Mechanisation brought the printing press, 'which was the first text "processor", the first technology of writing to duplicate words en masse' (Bolter, 1991:33). The printing press could produce multiple copies of books almost identical to the 'best' manuscripts. In the twentieth century, automatic typesetting speeded up this process. The computer then changed the technology of writing by adding flexibility to the rapidity and efficiency of generating text and printing. Electronic writing marks the next major shift in writing technology after the printed book.

Until it was computerised, writing was so much taken for granted that it was no longer recognisable as a technology. There was a time, however, when writing was a radical and innovative technology. Now computerisation has made the technology itself once again highly visible, especially in the cases of word processing, desktop publishing and hypertext. In contrast with traditional text, electronic writing depends upon an emergent technology which is still subject to transformation; 'indeed, all indications are that accelerating change is an inherent characteristic of this technology. It may *never* stabilise' (Slatin, 1990:873).

Although the relentless hype about the Internet and the World Wide Web may lead people to fear otherwise, the growing presence of the computer does not necessarily signal the death of the printed book. The introduction of a new technology of writing does not automatically render older ones obsolete, mainly because no technology has ever proven adequate for all needs. For example, even though printing completely replaced handwriting in book production, it did not spell the end for handwriting. Rather, the boundaries between the two writing technologies blurred. Today pen and paper serve for notes and personal communications; word processing and typewriting are for texts considered not ready or appropriate for typesetting. It seems that typesetting, word processing and handwriting will continue to complement each other, at least for the immediate future. As McLuhan (1964:8) observed, 'the "content" of any medium is always another medium'. More advanced technologies incorporate those that come before. 'Writing contains speech, print contains writing, film contains both these media, as when a voice-over accompanies a montage of headlines. Hypermedia, the latest of McLuhan's *extensions of man*, unites sound, graphics, print and video' (Moulthrop, 1991a:66). The future of writing is not a linear progression in which new technologies usurp old ones. 'It is more likely to be a *recursion*, a widening gyre in which new forms merge and coevolve with their precursors' (ibid.).

Attachment to a particular writing technology, usually either the printed book or computer, can arouse passionate debate. Both computer detractors and computer enthusiasts offer predict-able sets of arguments. Detractors point out that printed text is inexpensive, portable and easy to read, especially in bed or at the beach; moreover, writing with a pen is more natural than writing with a computer. By contrast, electronic text requires a machine which is expensive, vulnerable to breakdown and energy-dependent; it appears on a screen that is difficult to read, cannot be enjoyed comfortably either in bed or at the beach, and is aesthetically unattractive and dehumanising. Computer enthusiasts claim that electronic texts offer speed and efficiency in production and consumption, and at low cost. Computers also search through texts rapidly and display them in different forms. Enthusiasts are careful to distinguish the experience of reading electronic text from the particular technology on which it is read. Until recently, they agree, the computer has been a less appealing reading site than the printed book. But several new developments—such as the compact laptop computer and the wireless modem—promise to change the nature and location of computer use. Further, enthusiasts contend that writing with pen is no less technological and no more 'natural' than writing at a computer screen. Such arguments illustrate popular views of the differences between the two technologies. The following sections in this chapter explicate some of the more substantive distinctions.

The electronic writing space

The computer offers a medium for writing—what Bolter (1991:10) calls a 'writing space'—quite different from anything that has preceded it. In this 'late age of print' (ibid.:1), writers and readers still conceive of text as something located in a printed book. In the conceptual space of a printed book, writing is fixed and controlled by the author: that space is defined by bound volumes that sometimes exist in thousands of identical copies. The new writing space includes both the computer screen where text is displayed and the electronic memory in which it is stored. This electronic environment is characterised by fluidity, and enables an interactive relationship between writer and reader.

Electronic writing derives its effects from 'the central fact that computing stores information in the form of electronic code rather than in the form of physical marks on a physical surface' (Delany and Landow, 1993:7). Although the letters on the screen look like those on pages, they are in fact 'the temporary, transient representations of digital codes stored in a computer's memory' (ibid.). In other words, texts on the screen are 'virtual' in the sense that they are perceived to be different from what they really are.

Unlike the printed text, which can be held, the electronically produced virtual text is abstract; it is always 'a simulacrum for which no physical instantiation exists' (Landow, 1994:6). 'Virtuality' represents a new mode of existence that is neither actual nor imaginary but a simulated existence resulting from computation. Computer users cannot physically touch this space or move around within it. Nevertheless, the place in which interactions between readers, writers and text occur can still be defined as three-dimensional, since its length, depth and height can be explored by means of software tools.

The text encountered on the computer screen exists as a transient version created by the writers; an electronic primary version of it resides in the computer's memory. Writers work on an electronic copy until both versions converge when the computer is commanded to 'save' the current version of the text by placing it in the memory. At this point, the text on the screen coincides briefly with the text in the computer's memory. What the writer always encounters, however, is a virtual image of the stored text rather than the original version. The effect of these processes is to remove or abstract individual writers from their texts. The most unusual feature of electronic writing is that it is 'not directly accessible either to the writer or to the reader . . . There are so many levels of deferral that the reader or writer is hard put to identify the text at all: is it on the screen, in the transistor memory, or on the disk?' (Bolter, 1991:42–3).

'Text on the screen', writes Joyce (1992:86), 'is conspicuously non-print, unheavy, undark, dry, unimprinted, prone to sailing off'. Electronic text is dynamic and volatile:

> it somehow stays where it is when you scroll it, though it is always momentarily lost, or the eye is. It always threatens to go away. Even when static, electronic text seems frantic. It constantly reenacts its making, bobbing up from the dark beneath with the same buoyant insistence it has when you type it. Blurts of light emerge, more linear than ever print is in their constant inscription, the text always pushing the cursor forward, or dancing above a block of light like singalong lyrics in old movie houses. (ibid.:87)

The essential difference is that whereas 'print stays itself, electronic text replaces itself' (ibid.).

Working with virtual texts avoids the transformation of ideas into a graphic representation. The trace on the screen, Poster (1990:111) suggests, 'is similar in its spatial fragility and temporal simultaneity to the contents of the mind or to the spoken word'. Writers who begin to compose with computers find that many aspects of the writing process become much easier. The powers of the machine seem to resemble those of the brain: the technical feats of moving paragraphs from one position to another, merging two texts, and checking spelling or using a thesaurus appear to equal 'the achievements and capabilities of the mind' (ibid.:112). As with any technology of writing, it becomes difficult for the writer to determine where thinking ends and writing begins, or where the mind ends and the computer begins. The writer may come to regard the computer, and in particular its screen, as a metaphor for the mind.

Writing with a computer not only blurs the line between thinking and writing but also shapes to some extent the ways in which we think. The space created by each writing technology permits certain kinds of thinking and discourages others. A blackboard encourages repeated modification, erasure, casual thinking and spontaneity. A pen and paper invite care, attention to grammar, tidiness and controlled thinking. A printed page solicits rewritten drafts, proofing, introspection, editing. The computer, particularly when running hypertext-based applications, stimulates yet another way of thinking: 'telegraphic, modular, nonlinear, malleable, cooperative' (Kelly, 1994:463). We organise our writing space in the way we organise our thoughts, and in the way in which we think the world itself must be organised. The space of knowledge in ancient times was a dynamic, oral tradition. By the grammar of rhetoric, 'knowledge was structured as poetry and dialogue— subject to interruption, questioning, and parenthetical diversions' (ibid.:463–4). The space of early writing was similarly flexible.

Texts were ongoing affairs, amended by readers, revised by disciples; a forum for discussions. When scripts moved to the printed page, the ideas they represented became monumental and fixed. Gone was the role of the reader in forming the text. The unalterable progression of ideas across pages in a book gave the work an impressive authority. (ibid.:464)

The microcomputer provides a new writing space, which in turn has its own impact on the ways in which we come to know and to organise our thoughts.

The ways in which we construct knowledge are also closely connected to the metaphors we use or are exposed to. Some are associated with the structure, terminology and use of electronic text. Ideally, metaphors clarify and conceptualise the unfamiliar: affecting the ways in which we perceive, think and act, they are crucial to our understanding of any subject matter. An ineptly applied metaphor can result in confusion. A ubiquitous metaphor in electronic text is that of the traditional writing space: a desktop and related images of files and documents. Miniature desktops, little file icons, partially overlaid open files, pop-up notebooks, clipboards, rubbish bins for discards: it is as if the user had never left the familiar territory of paper. These business and desktop metaphors are at once alluring (in that they present a familiar frame of reference to explain something whose attraction is based in part on difference) and inadequate, because they are con-strained by metaphors of print textuality. They do not help us move beyond a print-view of the world towards understanding those new kinds of textuality and ways of knowing presented by the electronic medium. Current on-line documents demonstrate that the computing industry has either ignored or deliberately minimised 'the difference between paper and digital text, through metaphors and illusions' (Carlson, 1991:73). But if hypertext becomes the medium of the future, electronic text will be freed from the page, and the process of devising new metaphors will accelerate.

All electronic writing systems provide visual elements not present in a printed work. The most fundamental is the cursor, the blinking arrow, line or other graphic element which the user moves either from the keypad by pressing arrow-marked keys or with devices like a mouse or trackball. The cursor provides a mobile image of the writer's presence in the text. Because its positioning does not interfere with the text, its effect is substantially different from that produced by a reader moving a finger across a printed page or marking it with a pencil.

Whereas the page is the framing device in a printed book, in the current generation of machines the window is 'the defining feature of computer typography' (Bolter, 1991:69). The computer window not only marks out a space for a particular unit of verbal text, graphics or both, but also frames the writer's view of that space, which is 'an indefinite two-dimensional plane' (ibid.). In some systems, windows can be 'tiled', that is, 'set side by side so that the writer/reader can look onto two or more planes at once' (ibid.). In other systems, the windows can be 'stacked', so that 'the planes of text and graphics pile on top of each other' (ibid.). This is a different procedure from that of the printed book, where pages stay in the one consecutive order, each page hiding all the others underneath it. In the electronic medium, writers can move one window aside in order to view parts of the window below; they can reorder the stack 'by plucking one window from below and placing it on top' (ibid.). If the windows contain different texts, writers can move between them, adding to and cutting from each. 'No previous technology has offered anything quite like the windowed typography of the current microcomputer' (ibid.). Switching between windows is comparable to shuffling papers in a notebook, but nothing in print technology corresponds to the facility for enlarging the size of a window or scrolling through the text. Both window and text may change at any time.

It is useful to think of the differences between the printed page and the computer window, or the computer screen which displays the window, as reflecting different 'grammars' (Selfe, 1989:3). The computer screen represents part of the grammar of computers. Readers and writers are accustomed to the grammar or conventions of the page and the book, but are less familiar with the grammars of the keyboard, screen, software and network systems. We know that in cultures using European languages 'a page is read by starting in the upper left-hand corner and proceeding to the lower right-hand corner in a line-by-line fashion' (ibid.:5). Pages may exhibit such formal conventions as running heads, folios or numbers, and standard margins. They also have physical conventions: made of paper and bound within a book, they are a fixed size, arranged in a consecutive order, and constitute structural units of text that do not change from reading to reading. Pages are 'immutable' (ibid.).

Although the grammar of the computer screen shares some conventions with the printed page, the two are quite different. Pages are static units in a spatial text. Screens, by contrast, do not display units of text because 'they are temporal windows on a virtual text' (ibid.:7). The screen is therefore a temporal rather than a structural unit. Readers of screen texts are denied some of the spatial–contextual cues to which readers of page texts have access. 'A reader of a book', explains Selfe,

> can gauge its length in one glance and get an idea of format, organisation, and arrangement by flipping through pages. In contrast, the reader of a virtual text on a collection of screens can see the whole text only in his or her mind, and must keep track of individual screens in the same way. Given this lack of spatial-contextual cues, movement through a virtual text is often considered more difficult than movement through a printed text, either slower (when scrolling is done line by line), or more erratic (when one screen dissolves instantly into another at the touch of a key). (ibid.)

Screens also have different formal conventions. Unlike pages, they are not numbered, their margins are fluid, and they have cursors, windows and menu lines.

Selfe uses the metaphor of 'multilayered grammars' (ibid.:3) to express the sense of how writers trained in print adjust to on-line literacy. 'Each one of these grammars represents new layers of complex conventions that individuals must learn if they hope to function successfully and literately within such an environment' (ibid.:11). As print literacy is likely to remain the first language for some time, writers have to make efforts to learn the grammar of the screen—not just to move quickly from place to place, but also to use specific screen characteristics such as colour to indicate different levels of textual importance. Although word processors and electronic photocomposition are still being used to improve the production of traditional pages in books and typed documents, the page is not a meaningful unit in electronic writing. As Bolter (1991:3) suggests, electronic text 'must instead have a shape appropriate to the computer's capacity to structure and present' it.

Somewhat hyperbolically, Lanham (1993:73) hails 'the coming of the electronic word, the movement from letters printed on paper to digitised images projected onto the phosphorous screen of a computer' as a momentous transformation of literary studies. He identifies a number of basic changes: the fixed printed surface becomes 'volatile and interactive' (ibid.), while the immutable text upon which our culture has been built is called into question.

> We have come to regard print as so inevitable that we have ceased to notice its extraordinary stylisation. Print, after all, is a trickery too, not a historical inevitability. Print represents a decision of severe abstraction and subtraction. All non-linear signals are filtered out; colour is banned for serious texts; typographical constants are rigorously enforced; sound is proscribed; even the tactility of visual elaboration is outlawed. (ibid.:73–4)

The computer confronts us with a writing space which is different from anything that has preceded it. Indeed, we can now talk of a new mode of writing: electronic writing.

Computer as word processor

Technologies of writing strongly influence the ways in which writers write. Everybody who has converted to word processing has a story to tell about how the move has affected their writing processes. Freed from the need to retype everything, I have become more experimental and risk-taking. Professional writers notice differences in the pace of their writing, in the attention they pay to the relation between detail and large structures, and in their patterns of revision. Some writers treat word processors as instruments by which writing may be stored, slightly amended and printed. Word processing provides others, however, with the means by which thoughts, ideas and concepts can be formed, shaped and developed more flexibly than with pen and paper.

Word processing remains the most commonly used computer-writing application in business and academic settings. Computers equipped with word-processing software control the creation, storage, retrieval, modification and presentation of text. The program's ability to insert new material anywhere into a piece of writing makes it not much more complicated to revise a text than to leave it be. This feature alone is enough to highlight the important changes to writing practices brought about by word processing. Other important attributes include insertion of new material at the point of the cursor; the easy generation, manipulation and revision of text; the fact that it has to be entered only once; and the professional printed product at the end (Snyder, 1994).

Word processing makes a number of effective writing strategies substantially easier. For example, 'the ability to move into a working draft text from any number of "prewriting" activities (free-writing, research notes, outlines) makes turning early thoughts into writing a more useful enterprise' (Balestri, 1988:15). The fact that text has to be keyed in only once results theoretically in more time for writing and rewriting. Further, if composing is seen as the writer's search for meaning in an evolving text, 'then the word processor's greatest strength may lie in its capacity to aid in the unique unfolding of that text'. The computer makes it possible for writers 'to compose more freely' as it provides 'a flexible environment for research and writing' (Catano, 1985:312). It also offers writers a fluid context in which 'to creatively combine, rearrange and revise . . . ideas' (ibid.).

With word processing, identifiable phases of the writing process blend in a peculiar fashion. Well before computers appeared in educational settings, writing theory and research had revealed that planning, composing and revising are not necessarily discrete stages to be worked through in a linear direction; on the contrary, writers move in and out of each in complex, recursive patterns. Word processing simplifies the writer's capacity to respond recursively to writing problems and decisions.

When working with traditional tools, experienced writers devise ways of circumventing the constraints of the technology. Both pen and typewriter enable them to leave blank spaces or to skip lines. A typewriter, however, is even less flexible than a pen, and gives writers correspondingly less control at the level of both the sentence and the text as a whole. Even to move back and forth is a clumsy operation, and substantial revision usually demands a retyping of the complete text. In the case of revising both the handwritten and the typed page, the laborious procedure of cutting and pasting is only partially helpful in maintaining some flexibility. It could be argued that a text pieced together in this way 'is more fixed than ever, both literally and psychologically, by all that tape' (ibid.).

Writers new to word processing assume that here too they will be constrained, and that they will have to concentrate on the screen to the detriment of the text as a whole. But the screen text is fluid in the sense of being both a text in process and the process itself. 'The fluid text, to extend the metaphor, can always be allowed to rise and fall; the author can always be floating in new materials, leaving others behind' (ibid.). This breaks with the linearity of text production: the writer is not constrained by the end of either the line or the page. The text produced in this unfettered movement is itself as fluid as thought. It can be expanded or contracted, and its parts split, joined, reserved, discarded or inserted.

The fluidity of word processing transforms not only the content but also the appearance of the text. By allowing writers to consider every aspect of presentation, it enables the text to be reshaped for varying purposes and audiences, and to look like a published document. Writers can choose from a variety of fonts and formats, and at any stage of the writing process can see the format the words will assume in print. The visual aspects of composing can be manipulated: page layout and design—columns, margins, tabs, fonts, pitch and point size. No longer just a distraction from what has long been considered the more serious uses of written language to express and develop complex ideas, the graphic dimensions of writing themselves become important.

The reactions of writers to the use of computers suggest that even basic word-processing activities such as entering and manipulating text affect how they think and feel about composing. The computer's physical properties—such as kinetic screen display and immediate responsiveness—are especially significant. Writers who compose with computers are required to conceive of 'text' in new ways. Accustomed to the page, notecards, books and filing systems, they have to adjust to a text that can be displayed on a screen, recorded in the computer's memory, stored on a disk, and controlled and generated by commands given to the machine. Although it is recorded in an inaccessible place inside the machine, the stored text becomes strangely tangible as an entity that can be moved about, put away and brought out again to be re-examined. The writer looking at a two-dimensional screen may be imagining a much larger and three-dimensional world, which includes the rest of the text off-screen at that time.

Although Bolter (1992a) admits that word processing offers new flexibility in allowing writers to copy, compare and discard text with the touch of a few keys, he argues that in the final analysis it does not challenge conventional notions of writing. The word processor is merely a means of perfecting printed or typed copy: the goal is still ink on paper. It is therefore not revolutionary because it does not force us to reconceive text by providing 'a new space for writing and reading: instead, it gives us a space that is an electronic image of the familiar typed or printed page' (ibid.:19). In Bolter's view, the computer as word processor is still an intermediate tool used to prepare texts that eventually will be translated into the older medium of ink and paper. It marks the transition between conventionally print-directed writing and that fully electronic writing which is hyper-text.

However, it is probably a mistake to assume that a new technology will be used simply to extend, rather than to alter, an existing practice. New technologies designed merely to streamline or otherwise enhance the conduct of familiar social routines may end up so reorganising them that they become new events. Word processors make the business of writing seem like a different activity. As Balestri (1988:14) observes, even a print-based technology like word processing changes the habits of writers 'in unique and unpredictable ways'. Writing with a word processor is different from writing with a pen or a typewriter, and therefore more than just a new way of doing old things.

From softcopy to hypertext

Writers who aim at producing neatly presented printed versions of their texts are composing in what Balestri (1988) calls 'hardcopy'. Still 'writing to the printer' (ibid.:15), they are using the computer in a limited way, treating it as an electronic typewriter which mechanically makes text, drives a printer, and generates hardcopy. Such writers miss the opportunity presented by the computer to move beyond word *producing* to word *processing*.

When writers forget the printer and write directly to the screen they are composing in what Balestri calls 'softcopy' (ibid.:16), which is the text on the computer screen. When working in softcopy, writers 'experience a new kind of hand-work that, while removing some of the intimacy of hand-wrought notation and symbol, actually improves on the non-linear capabilities of handwriting' (ibid.). To write to the screen is to be unconcerned with a printed product that may or may not eventuate. Softcopy offers a process-oriented rather than a product-oriented mode of writing. Composed in softcopy, writing is no longer a fixed inscription but instead a fluid and malleable stream of electronic information. Softcopy documents are written to be displayed rather than printed, and designed for provisional recording in electronic storage, pending the rereading or rewriting of them.

Balestri's distinction between hardcopy and softcopy usefully articulates the shift from print-oriented computer applications (such as word processing) to hypertext. It explains the difference between that static and product-centred conception of writing which still dominates academia and schools, and the more dynamic and process-centred conception of electronic discourse that is beginning to influence educational theory and practice. Hypertext is a wholly electronic form of writing that uses the computer as a medium in its own right, both for the creation and the reading of texts. Hypertext enables us to stop thinking of the text as a series of printed pages whose components are ordered immutably, paragraph by paragraph, from first to last:

> The computer allows us to define units of text of any size and to present those units in a variety of orders, depending upon

the needs and wishes of the reader. An electronic text is fluid, adjustable right up to the moment of reading. Indeed an electronic text only exists in the act of reading—in the interaction between the reader and textual structure. (Bolter, 1992a:20)

Hypertext thus allows the computer to 'display in full measure its unique qualities as a writing space' (ibid.).

Explaining Hypertext

The architecture of hypertext

Imagine you are reading a standard scholarly article on Shirley Hazzard's novel *The Transit of Venus* (1980). As you read through the main text, you encounter a symbol that indicates the presence of a foot- or endnote. You then leave the main text to read the note which may refer you to a commentary by another author, information about parallels with earlier or later works by Hazzard or, perhaps, material about the influence on her fiction of other literary texts. You can either track down these other texts or return to the main one and continue reading till you encounter a further note, whereupon the same process will be repeated. This kind of reading, familiar to all of us, 'constitutes the basic experience and starting point of hypertext' (Landow, 1992a:5).

Suppose now that by simply touching the page you could bring into view instantly one of those other texts, such as *The Evening of the Holiday* (1966), an entire novel by Hazzard. And if you were to touch the page again, you would encounter another text. But as we know, this is not possible, because in print technology the referenced materials 'lie spatially distant from the references to them' (Landow, 1992a:5). By contrast, hypertext simplifies the following up of references: the field of interconnections is easily traversed and the linking is instantaneous. In hypertext, a note can be longer than the work it annotates: it may be another whole work, or be linked to other notes or annotations that are themselves complete texts. One's experience of hypertext is not so much non-linear as multi-linear or multi-sequential. And it is not just in literary studies that hypertext has potential. With a hypertext system, a historian might choose to write an essay in which each paragraph or section is a unit or 'text node' in a hypertextual network. The various interconnections would constitute possible orders in which such text nodes could be assembled and read, and each order would produce a different outcome. Hypertext also has more popular uses in the compilation of directories, catalogues, dictionaries, how-to manuals or any other text through which readers normally move in an order of their own choosing.

The computer's capacity to create such fluid structures and present them interactively to the reader results in hypertext, which is more than automated footnotes. Unlike footnotes in a printed book, the process of reference can continue indefinitely in the computer. In a book it would be intolerable to encounter footnotes to footnotes. But in the computer, writing in layers is both possible and tolerable, and reading a multi-layered text is effortless. Each constituent text may be equally important in the whole, which then becomes a network of interconnected writings. Effective reading can be done only at a computer screen, because only computers can activate the links and take the reader effortlessly from one text to another.

Like a scholarly article, encyclopaedias and magazines are also forms of non-electronic hypertext, collections of writings through which readers are free to move in almost any sequence. Unlike an encyclopaedia, however, a hypertext does not present its readers with a predefined structure. 'The "articles" in a hypertext are not arranged by title or subject; instead each passage contains links or reference markers that point toward other passages' (Moulthrop, 1989:18). These markers may be actual words in the text, keywords implied by the text, or special symbols. To activate the link, usually by means of a pointing device or mouse, is to bring the indicated passage to the screen. The major difference between the structure of a scholarly article or encyclopaedia and that of hypertext is the creator's ability to make various links automatically, and to cross from one medium of communication to another. Moreover, the body of written and pictorial material is interconnected in such complex ways that it could not be presented conveniently on paper.

Hypertext disturbs our linear notion of texts by disrupting conventional structures and expectations associated with the medium of print. With some notable exceptions, books are essentially repositories for the sequential storage of information. And although contemporary reading and writing theorists argue that readers do not simply progress word by word, line by line and page by page until they have 'finished' the text, this conception of reading is nevertheless common. Information in books is stored according to unchanging spatial grids, in so far as the same information is presented in the same order on the same page every time a book is opened and used. Unlike books, however, a hypertext need have no beginning, no immutable order in which the information is set out, and no ending. It may offer multiple entry points and provide many different pathways through the system for readers who themselves choose when and where to leave it. Unconstrained by the linearity and arbitrary ordering of the print medium's presentation of information, hypertext users can browse through the data more freely.

By enabling its readers to construct hybrid documents based on associational links rather than linear sequences, hypertext departs from the print-oriented technologies of word processing and desktop publishing. In hypertext, the reader moves from one text node to another, either by following established links or by creating new ones. A print-bound text is the result of many individual choices made by its author from among several available alternatives. By contrast, a hypertext consists of many virtual texts which may be the work of different writers. Each reader makes one or more of these virtual texts an actual text when choosing which links to follow and which to ignore.

In hypertext, the window (which is the defining feature of computer writing and reading), takes on 'a structural significance' (Bolter, 1991:70). Operational links enable texts in one window to be linked with texts in another. 'Following a link can make windows appear, disappear, or rearrange themselves so that the destination text comes to the front of the screen and captures the reader's attention' (ibid.). In addition to the cursor and the linking symbols, hypertext systems use various devices to orient navigators through hyperspace, such as graphic overviews, 'concept maps', directories and 'web views' which provide readers with a sense of either the whole web or which documents surround the one they are currently reading. The most common is the 'concept map', which informs readers about linked material and provides clear access to it. Such overviews organise efficiently a body of complex ideas whose centre will vary in the opinion of different readers and writers.

Because of such graphical orientation devices, the visual appearance of the text assumes a new status in hypertext systems. By integrating the currently separate worlds of pictures and words, hypertext exposes our western cultural bias towards information which can be measured by pages and paragraphs comprised of words. Writers have internalised the belief that verbal information is more valuable than non-verbal information, and that non-verbal elements are the business of publishers, designers and printers, not of writers. Much more than word processing, however, hypertext demands that writers pay careful attention to the non-verbal. Since we are relatively sophisticated in our understanding of such realistic representations of information as photographs and film clips, we do not have much difficulty with these media forms. We have, however, 'a generally naive understanding of graphics' (Carlson, 1991:87), which figure significantly in hypertext. Learning how to read, produce and exploit graphics constitutes one of a number of new demands imposed on users by the technology.

Hypertext is essentially a network of links between words, ideas and sources that has neither a centre nor an end. We 'read' hypertext by navigating through it, taking detours to notes, and to notes to those notes, exploring what in print culture would be described as 'digressions' as long and complex as the 'main' text. Any other document can be linked to and become part of another text. 'Computerised hypertext incorporates marginalia and commentaries to the text by other writers, updates, revisions, abstracts, digests, misinterpretations, and as in citation indexing, all bibliographic references to the work' (Kelly, 1994:462). The extent of hypertext is unknowable because it lacks clear boundaries and is often multi-authored.

Such features appear to constitute the generic characteristics of hypertext, although of course it is just as difficult to talk of 'generic' hypertext as of generic print. Not all printed texts appear in books, for instance, nor for that matter as literature. Similarly, hypertext systems are not used exclusively for scholarly purposes in cultural and literary studies: in fact, they are used more widely by 'knowledge workers' (Landow, 1994:8) in applications such as aeroplane-repair manuals. As with print, the projected audience for hypertext largely determines the system's characteristics. Instead of trying to discern its generic qualities, therefore, it is more helpful to identify the principal types of hypertext. For example, stand-alone hypertext systems such as Hypercard (which constitute most people's experience of hypertext), differ from networked or distributed hypertext systems such as the World Wide Web, in which 'digitised text and other data can be manipulated simultaneously by many users' (ibid.:9). We can also differentiate between 'read-only' hypertext like CD-ROMs (in which readers' contributions are limited to choosing their own reading paths) and hypertext in which users can add text, links or both.

Origins of hypertext

The creation myth of hypertext is dominated by the rhetoric of founding fathers and pioneers. Although appealing in its patriarchal simplicity, romanticism and elegance, this approach to history has its limitations. Equally problematic is the impulse to explain the history of hypertext technology—indeed of anything—as a neat series of chronological and causally linked events. Histories of hypertext presented in this way may be comforting in their formal familiarity, and apparently illuminating, but they remain merely constructs. Acknowledging these difficulties, I attempt to provide in the following sections an overview of the contributions made by key individuals to the development of hypertext, and to give a sense of the socio-cultural and literary formations which produced them. The decision to begin this chapter with an expanded explanation of hypertext reflects my desire to resist the impulse to represent the development of the technology as a relentlessly linear progression. More importantly, such an approach enables us to understand what early thinkers in the development of hypertext were working toward.

For centuries scribes, scholars, publishers and other makers of books have been inventing devices to increase the speed of information retrieval. 'Manuscript culture gradually saw the invention of individual pages, chapters, paragraphing, and spaces between words. The technology of the book found enhancement by pagination, indices, and bibliographies' (Landow, 1992a:19). All of these devices have facilitated reading. Yet although multiple copies of the fixed texts made possible by print technology have had enormous effects on modern conceptions of literature, education and research, they nevertheless make information-retrieval difficult because they preserve their information in an unchangeable linear format which is totally different from the dynamic, alterable and multi-sequential format of hypertext.

The problems of linear text prompted some innovators to think of alternative ways of presenting information even before computers themselves became a reality. Commonly recognised antecedents of computerised hypertext include:

- Samuel Taylor Coleridge's 'Treatise on Method' (1849), which outlines the principles for organising all human knowledge
- Vannevar Bush's *Atlantic Monthly* essay, 'As We May Think' (1945), which is recognised now as the first serious attempt to lay out the principles and functions of a memory machine (the memex)
- Ted Nelson's (1978) vision of electronic hypertext, which he named Project Xanadu.

Coleridge, Bush and Nelson were all concerned with the problem of creating a system for providing complete access to the 'endlessly expansive world of texts' (Tuman, 1992a:55). The prototype of a hypertext system—NLS (oNLineSystem)—was first designed and built by Douglas Engelbart in 1968 at the Science Research Institute at Stanford University. What Bush and Nelson could only imagine, Engelbart made a reality.

Encyclopaedias

Coleridge intended his 'Treatise on Method' to be the introduction to a proposed encyclopaedia called the *Encyclopaedia Metropolitana* (1849). Having an intense dislike of alphabetical systems of ordering knowledge, he was seeking a set of principles that had more meaning than 'an arrangement determined by the accident of initial letters' (Collison, 1966:231). His 'Treatise' accordingly outlines alternative principles for organising all human knowledge, a problem of increasing interest nowadays, 'given the huge database capacity of modern computers and . . . the seemingly geometrical increases in information' (Tuman, 1992a:53).

While Coleridge's topical arrangement can be traced back to the Middle Ages, his ideal encyclopaedia is 'a clear product of the technology of print, in which the text is laid out in one ideal order' (Bolter, 1991:92). The ordering principle favoured by Coleridge is to demonstrate how each notion is subordinated 'to a preconceived *universal* Idea' (Coleridge, 1849:22): in other words, to present hierarchies of knowledge. 'The passion for hierarchy', argues Bolter (1991:105),

> finds its purest expression in the elaborate table of contents of modern encyclopaedias and other great books in print. The table of contents is both hierarchical and linear: it shows subordination and superordination, and it also shows the reader the order in which he or she will encounter these ideas in reading from the first page to the last.

In a hypertext nothing corresponds to the printed table of contents. Menus can indicate a hierarchy of topics, but the order of pages does not compel readers to move linearly through the structure. Hypertextual relationships are correspondingly multiple and evolving. Bolter (ibid.) goes so far as to suggest that hypertext is a writing technology well suited to the contemporary view that nature is not a hierarchy but 'a network of interdependent species and systems'.

Encyclopaedias have been the traditional print attempts to cover all knowledge, and it is perhaps not surprising that we now have encyclopaedias in electronic form. The literal precursor was the fifteenth edition of the *Encyclopaedia Britannica*: first issued in 1974, it was both an excellent printed encyclopaedia and 'a book straining to break free of the limitations of print' (ibid.:92). It had both a topical and an alphabetical arrangement: the main articles were printed alphabetically in volumes called the *Macropaedia*, to which separate volumes called the *Propaedia* provided a vast outline. In effect, the *Propaedia* 'turned the encyclopaedia into a hypertext whose parts could be assembled and reassembled by the reader' (ibid.:92). Because the references were hard to follow in a printed work of thirty volumes, the format was eventually abandoned. An index was added, and by the mid 1980s the *Britannica* had become again a conventional encyclopaedia.

Commercially available electronic encyclopaedias such as Grolier's (1988) are not 'true electronic books, but rather printed books that have been transferred to the computer' (ibid.:95). Even though readers can search for topics in a number of ways, they can neither intervene in the structure of the encyclopaedia nor build new structures. Indeed, most of the existing electronic encyclopaedias 'do not reflect the power or the limitations of the new medium, but rather the conservative character of the publishing industry, which is bound inevitably to the technology of print' (ibid.:95–7).

Bush's memex

The earliest conceptual framework for hypertext is generally believed to have been provided by Vannevar Bush, Director of Scientific Research and Development for the Roosevelt administration during World War II. In an essay in 1945, Bush envisioned a memory machine, which he called the memex, to manage the large volume of scientific information available at the time. Whereas Coleridge pointed to the limitations of alphabetic ordering, Bush traced the main problems of information retrieval to inadequate indexing and categorising systems. He imagined a machine that would transcend the storage and retrieval limitations of print technology by allowing users to gain access to and to search huge amounts of information in order to retrieve and annotate what they considered important. His memex would mechanise a more efficient and more human mode of manipulating information. Like the imagination, it would operate by association, and therefore better accommodate the way the mind works. Memex users would be able to browse through information by creating 'numerous trails' (Bush, 1945:104) of their own associative links. Bush describes how records of these memex trails could be stored and retrieved for another purpose at a later date.

He recognised that trails of links would constitute a new form of textuality and a new form of writing. 'When numerous items have been thus joined together to form a trail', he explains, 'it is exactly as though the physical items had been gathered together from widely separated sources and bound together to form a new book' (ibid.). Bush's conception of textuality introduced three new elements: associative indexing (or links), trails of such links, and webs of such trails. These elements 'in turn produce the conception of a flexible, customisable text, one that is open—and perhaps vulnerable—to the demands of each reader' (Landow, 1992a:17). They also introduce the idea of 'multiple textuality' (ibid.). For the memex world, 'texts' are not only those individual reading units that make up a work, but also entire works and sets of documents created by the trails, and perhaps even the trails themselves, without accompanying documents.

The memex machine would also allow users to add their own marginal notes and comments. In recognising the need to record one's reactions to texts, Bush conceives of reading as an active process that involves writing. His memex enabled the practices of reading and writing to draw closer together than is possible in book technology. However, the principal capacity of the memex was neither retrieval nor annotation but associative indexing, 'the basic ideas of which is a provision whereby any item may be caused at will to select immediately and automatically another' (Bush, 1945:103).

Writing in the days before digital computing, Bush conceived of his device as a desk with translucent screens, levers and motors for the rapid searching of microfilm records. A computer is indeed a perfect memex machine, able to handle large amounts of information, to display it on a screen, and to provide the means of creating trails that link relevant bits of information. But it can offer further benefits. For example, word processing enables the text not just to be read, but to be changed or annotated. The computer can search for words and phrases, reach distant sources of information over a network, and combine pictures and sound with text.

Nelson's vision

Like Coleridge and Bush before him, Ted Nelson was also concerned with how best to structure information so that it may be retrieved easily. Drawing on the speculations of Bush and others, in the early 1970s Nelson argued that what we need is not an ever-expanding collection of books but a system that gives users access to the total world of human knowledge, which he called the 'docuverse', and which is 'not unlike the linked world (as) text of contemporary computer networks' (Joyce, 1995b:23). Nelson called his docuverse project 'Xanadu': it is a system in which the whole of recorded discourse—all the world's 'literature', defined as 'an ongoing system of interconnecting documents' (Nelson, 1992a:2/9) —would be woven into one enormous matrix. Xanadu represents a design 'for the universal storage of all interactive media, and indeed, all data; and for a growing network of storage stations which can, in principle, safely preserve much of the human heritage and at the same time make it far more accessible than it could have been before' (ibid.:0/6). Although Nelson describes Xanadu as an ongoing project, it could be argued that the concept has been realised already in the World Wide Web, a system (currently used in the academies and industry) which allows multiple, simultaneous and synchronous interaction between those who create or gather material and those who use it.

Integral to the Xanadu project is Nelson's notion of hypertext, the term he coined in the 1960s to refer to *'non-sequential writing*—text that branches and allows choices to the reader, [and is] best read at an interactive screen' (ibid.:0/2). Nelson's hypertext encompasses texts that are minimally or maximally non-linear and tightly or loosely structured. In the broadest sense, all texts are hypertexts: even a printed text gives readers and writers one link out of each node (usually a sentence or paragraph), namely the option to move linearly to the next section of the text. Nelson thought it important to find the right word for his invention:

> *Hypertext* was an audacious choice: *hyper-* has a bad odour in some fields and can suggest agitation and pathology, as it does in medicine and psychology. But in other sciences *hyper-* connotes extension and generality, as in the mathematical hyperspace, and this was the connotation I wanted to give the idea. (Nelson, 1992b:49)

Hypertext is Nelson's alternative to traditional texts which provide readers with only one path (namely the author's) through a given body of information. His system of organising materials enables readers to move through it polysequentially in pursuit of their own ends. By offering us multiple pathways and the ability to make our own connections, hypertext more closely reflects the fluid way in which we think. Like Bush, Nelson thinks the mind works by association: with one item in its grasp, it moves instantly to the next, thus forming an intricate web of trails. By representing the 'unbroken web' (Drexler, 1986:224) of human knowledge and problems, hypertext allows users to 'keep ideas hitched together in ways that better represent reality' (ibid.:222). In hypertext, everything is intertwined and intermingled with everything else; or as Nelson puts it, 'intertwingled' (1978:DM2).

Nelson's vision of hypertext materialised in the form of Engelbart's NLS (oNLineSystem), later renamed 'Augment'. In the process of conceiving his system, Engelbart also 'invented or first put to serious use fundamentals of computer interaction, writing, and networking, including word processing, outlining, windows, electronic mail, computer conferencing, collaborative authorship, and—not last—the mouse' (Joyce, 1995b:22). Hypertext software first became commercially available in the 1980s. Examples of hypertext systems include Brown University's Intermedia, Xerox Parc's Notecards, Owl International's Guide, MIT's Media Lab applications, and Bolter, Joyce and Smith's (1990) Storyspace, a vehicle for exploring non-linear narrative. In the late 1980s, when Apple Computer began including its Hypercard with all new Macintosh computers, 'hypertext achieved an unprecedented ubiquity —or at least its potential did' (DiPardo and DiPardo, 1990:8).

The Xanadu project continues to develop apart from these other advances in hypertext. It has two main components. A company called the Xanadu Operating Company was given the task of producing what Nelson calls 'a hypermedia server program' (Woolley, 1992:159). This was to provide the mechanism for exploring large computerised databases of information, comprising video, music and voice as well as text. The server was also to provide the means of creating new documents by making links between the content of existing ones, so that individual users would be able to establish their own redactions of the database's contents. 'These links are the key to the Xanadu concept, since it is through links, through the creation of structures within a vast, shapeless mass of information, that Xanadu creates new meanings and interpretations that would be inaccessible using conventional methods of information storage' (ibid.).

The second component of the Xanadu project is to create a new publishing market by establishing a network of databases. Data contributed by publishers would gradually accumulate into a global information repository. The hypermedia server program would enable users to explore this repository. For every bit of text they linked to, users would pay a royalty to its publisher, even if the particular document they compiled from their session in Xanadu was made up of texts from a variety of different publishers. This royalty-payments system is a key component of Xanadu, since it would 'create a market, a trade of texts that would encourage more publishers to contribute, which, in turn, would attract more users' (ibid.:160).

In fact, Nelson's Xanadu revolutionises the conventional library which users can only consult or borrow from. In Xanadu, every reader becomes a potential writer, because the system is specifically designed to make it as easy to contribute to a text as to consult one. 'It creates an open market, a free trade in knowledge, where the success of a text is simply dependent on the number of times it is accessed, which in turn will depend on its relationship with other texts on the system' (ibid.).

The unifying structure of the Xanadu project, explains Nelson, not only maintains the integrity of the original materials it contains but also allows users to quote from and anthologise them in any way they wish. This can be achieved by what Nelson (1992b:54–5) calls 'transclusion':

> When you cite something, you ordinarily insert a copy of the quoted material from the original, or quoted, document into the new, or quoting, document. In the Xanadu model we use transclusion instead: now you have a hidden pointer in the data structure of the second document, which points to the original and tells the computer-based reading machine where to get it. So the material is not copied from the original; it remains in the documentary space of the original and is brought anew from the original to each reader.

Nelson's (1992b) commitment to social and cultural change—a political position not uncommon in the passionate rhetoric associated with the development of hypertext—is expressed in his somewhat fanciful articulation of the purpose of Xanadu:

> Our objective at the Xanadu project has been not to fulfil the needs of industry, or to make things happen a little faster or more efficiently. Ours has been the only proper objective: to make a new world . . . Open hypertext publishing is the manifest destiny of free society. It is fair, it is powerful, and it is coming. (ibid.:56–7)

In formulating the notion of Xanadu, Nelson (ibid.:52) poses a rhetorical question:

> Is it mathematically possible to supply billions of readers at screens with the exact paragraph, sentence, fragment, illustration, or footnote, photograph, or piece of movie that each requires, immediately? Even if the number of the stored documents and the number of links between them grow into trillions?

Nelson continues to believe that it can be done:

> I have a vision for the year 2020; I like to call it the 20/20 vision. Think of everyone at screens: a billion around the planet. And each person at a screen will be able to extract from

a great common pool any fragment of whatever is published, with automatic royalty and no red tape. (ibid.:44)

While it may be argued that Nelson's docuverse is not feasible because 'there are insurmountable political and social obstacles to a universal system' (Bolter, 1991:103), his vision remains an important incentive to hypertext development, even though its utopian excesses make it for the time being unachievable.

Borges's library

The notion of hypertext originates in the imagination not only of scientists but also of literary visionaries. In a story called 'The Library of Babel', the Argentinian writer Jorge Luis Borges (1970b:81–2) imagines a library of incomprehensible immensity:

> its shelves register all the possible combinations of the twenty-odd orthographical symbols (a number which, though extremely vast, is not infinite): in other words, all that it is given to express, in all languages. Everything: the minutely detailed history of the future, the archangels' autobiographies, the faithful catalogue of the Library, thousands and thousands of false catalogues, the demonstration of the fallacy of those catalogues, the demonstration of the fallacy of the true catalogue, the Gnostic gospel of Basillides, the commentary on that gospel, the commentary on the commentary on that gospel, the true story of your death, the translation of every book in all languages, the interpolations of every book in all books.

'The Library of Babel' comprises 'an indefinite and perhaps infinite number of hexagonal galleries' (ibid.:78), each containing twenty shelves, five long shelves per side, each shelf containing thirty-five books, each book containing 410 pages, each page containing forty lines, each line containing eighty letters. No author's name or title is inscribed on any of these books; they are simply arbitrary collections of symbols, each one a combination of all the possible combinations of the letters of the alphabet. That is why the library is so vast and contains every possible text.

The narrator describes the crazy and often desperate reactions of the inhabitants as they come to realise the implications of living in a universal library of random typography which exhausts symbolic thought. Every combination of the letters of the alphabet has been realised, and now sits on the shelves waiting for readers. There is nothing left to be written, although much to be discovered; yet discovery is impossible because the nonsense books overwhelm those that are supposed to have meaning. The inhabitants of this world, whom Borges calls 'librarians', wander about the cubicles looking for sensible books, but they are helpless before the logic of permutation. Their library, writes Bolter (1991:138), is

> the ultimate static text; the frozen technique of the printed word has become the universe. The exhaustion of writing also means that time has stopped for these readers. The librarians exist in an eschatological moment in which there is nothing left to wait for, because nothing new can be described.

For Borges, literature is at a dead end, an impasse, because of its commitment to single story-lines, denouements and conclusive endings. The American novelist John Barth has characterised Borges's work as 'the literature of exhaustion' (Barth, 1967:29); with Borges, we sense that a print-based literary tradition is breaking down. To renew literature, writers 'would have to write multiply, in a way that embraced possibilities rather than closed them off' (Bolter, 1991:139). But 'Borges himself never had available to him an electronic space, in which the text can comprise a network of diverging, converging, and parallel times' (ibid.). He could not see that the 'exhaustion' of literature is merely the effect of a print-bound technology now surpassed by the electronic medium of hypertext.

Although there is no real Library of Babel, its books in some sense exist as part of the global information network. Indeed,

> it is, perhaps, possible to discover texts that no individual author has so far found, variations on existing texts, alternative endings to familiar scenarios, elucidations, glosses, diversions, a new text that can be discovered by its proximity to an

existing one. That, at least, is the hope of the champions of a new sort of literary object: the 'hypertext'. (Woolley, 1992:153)

Joyce's paradigm

Michael Joyce (1995c:41–2) identifies two broad categories of hypertext: exploratory and constructive.

> Exploratory hypertexts encourage and enable an audience (*users* and *readers* are inadequate terms here) to control the transformation of a body of information to meet its needs and interests. This transformation should include a capability to create, change, and recover particular encounters with the body of knowledge, maintaining these encounters as versions of the material, i.e., trails, paths, webs, notebooks, etc. . . . Constructive hypertexts require a capability to act: to create, change, and recover particular encounters within the developing body of knowledge. These encounters, like those in exploratory hypertexts, are maintained as trails, paths, webs, or notebooks, but they are versions of what they are becoming, a structure for what does not yet exist. Constructive hypertexts require visual representations of the knowledge they develop.

These definitions provide a convenient and useful framework for understanding the full range of forms that hypertext systems might possibly assume. In some ways, Joyce's model evokes a continuum. At one extreme, a hypertext document may be so restrictive that readers find they have no more (and perhaps even fewer) navigational choices than they would with a linear version of the text. At the other extreme, a hypertext document may be so open, interconnected and reader-controlled that users could be overwhelmed by the multiplicity of choices. To understand the differences between these two types of hypertext, consider the following hypothetical example of an exploratory hypertext.

You sit in front of a screen and open a Henry Handel Richardson web. The title *The Fortunes of Richard Mahony* appears in a menu surrounded by additional works by Richardson, along with other texts such as biographies, histories of Australian literature, maps of Victoria and the goldfields, clips from documentary films on the goldrush period, and contemporary poetry and art. You can browse through any of these texts, or alternatively open *The Fortunes of Richard Mahony* (1930). When you are reading the novel, you can pursue anything that interests you by placing the cursor on a highlighted word and clicking the mouse. You may contribute your own texts and add links to the web. As you explore the hypertext, you proceed in directions of your own choice.

An exploratory hypertext, then, is an interconnected world of texts on a particular subject. In moving around the hypertext, the reader is not bound by the conventions of linear narrative. Readers become explorers when invited to interact with a body of knowledge to meet their needs and interests. However, despite the appearance of 'interaction', exploratory hypertexts are perhaps just as 'autonomously meaningful' as books and 'like books they also preserve the discursive autonomy of their authors' (Moulthrop, 1991b:153). Although readers may 'transform' the textual body by following alternative pathways, it 'retains its fundamental identity under all transformations' (ibid.). The maze may have many permutations, but the users still move through 'the same mechanised volume, working toward some predefined end' (ibid.) established by its creators—even though reaching the 'end' is more like exhausting all the information provided in the system than experiencing the resolution and closure of a novel.

Whereas exploratory hypertexts are designed for audiences, constructive hypertexts are designed for writer-operators. It is difficult to simulate the experience of navigating a constructive hypertext in a traditional book. Writers use them to develop a body of information which they map according to their needs and interests; they transform the information into knowledge as they invent, gather and act upon it. The operators of a constructive hypertext have a much greater degree of freedom. 'Constructive hypertexts extend not just in material volume, but also into the deterritorialised space of electronic writing. They are not closed books but open ranges, discursive improvisations that grant no last word' (ibid.:154). Indeed, Joyce's notion of constructive hypertext is in accord with Nelson's explanation of the docuverse. Instead of being considered unitary documents, hypertexts can be conceived of as 'multiple discursive sequences' (ibid.). Because hypertext is non-sequential, it allows movement from the notion of a bound volume to a true 'writing space'. Electronic writing in the form of constructive hypertext enables an *'unbinding* of the text' (ibid.:153).

The shift in metaphors from 'exploration' to 'construction' is significant:

> The reader or 'user' of an exploratory hypertext can 'transform' the discourse by subjecting it to a range of manipulations predetermined by the text's designer. But the *operator* of or on a constructive hypertext . . . has a much greater degree of freedom . . . She is no longer an explorer, discovering what was left for her to find, but a constructor or co-creator, a belated but equal participant in an unfolding 'social' text. (ibid.:154)

Balestri's notion of softcopy—a fluid and malleable text in flux—is also relevant to Joyce's explanation of constructive hypertext. Balestri (1988) argues that word processing changes fundamentally the production of writing by replacing hardcopy with softcopy. Softcopy documents are infinite processes, not definitive products, and are always open to expansion and extension by other writers as well as by their originators. In this respect they resemble Joyce's notion of constructive hypertexts which are undefined, expansive models of developing conceptual structures. Those hypertexts Joyce calls 'exploratory' are the closest thing to hardcopy in the electronic domain, and their value is correspondingly more limited. Constructive hypertext, by contrast, represents 'a form of writing that does not attempt to duplicate the functionality of print but instead sets out to explore new possibilities for written communication' (Moulthrop, 1991b:154).

Both Balestri and Joyce suggest that 'our new understanding of textual authority must be centred not on singularity, consistency, and closure, but on difference, multiplicity and community' (Moulthrop, 1991c:267). They argue that 'we must reconceive writing not as a private activity eventuating in a public product, but as a process of revision or construction that is itself shared between writers and readers, or among reader/writers' (ibid.). As such, hypertext offers opportunities to alter the nature of the academy and the kinds of writing practices traditionally located there. Like many other writers on hypertext, Balestri and Joyce use an idealistic rhetoric of 'community' and the 'democratisation of the academy' (examined critically in Chapter 6).

At this stage in the history of the technology, most extant hypertexts merely reproduce existing works in the new medium. They are, in Joyce's term, 'exploratory'. In such works, the hypertext component is limited to navigational devices that facilitate exploration of an information space. The users remain in 'audience' mode and the roles of the readers and authors remain separate and different. It is not 'exploratory' but 'constructive' hypertexts that offer radical alternatives to educational processes. Indeed, Joyce's choice of the word 'constructive' has a familiar resonance for those working in education, where 'constructivism' is regarded as a particularly powerful theory of learning. As with constructivist approaches to teaching and learning, 'knowledge' in constructive hypertext theory is seen as existing not as a preconceived truth waiting to be discovered, but rather as a potential: until we create it, link it, write it or recover it, it does not exist. 'We create this knowledge contextually and share it electronically not by convincing someone that we are right, but by following their exploration of our links and exploring theirs in order to negotiate our shared and disparate spaces' (McDaid, 1991:214–15).

Navigation and orientation

A problem common to both exploratory and constructive hypertexts is 'navigation', a metaphor used to describe how users move through hypertext documents. Any system that links potentially vast amounts of knowledge could make navigation difficult by overloading its users with choices. Users need to proceed from one place to another without getting lost along the way. Bush (1945:105) suggested a solution to this problem by introducing the notion of 'trail blazers', that is, human navigators who would combine the skills of intellectual historians, biographers, psychologists and archivists. Using Bush's memex system, they would record the trails by which discoverers came to their findings.

It is certainly easy to get lost in a hypertext document. Lacking the centring hold of a narrative, everything in the document seems to have equal weight, wherever the user goes. The problem of locating items in the web may be substantial: 'the lack of simple but psychologically vital clues, such as knowing how much of the total you've read or roughly how many ways it can be read, is debilitating' (Kelly, 1994:463). Writers need to develop navigational aids and representational structures that will diminish the possibility of becoming 'lost in hyperspace' (Dede and Palumbo, 1991:15). Unless the problems of navigation and filtering are handled carefully, 'even the most liberal and benign hypertext systems can begin to resemble bureaucratic and even oppressive structures' (Moulthrop, 1989:24).

Anticipating the problem that users might possibly become disoriented in a hypertext environment, Landow (1991:82) emphasises the utility and intentional coherence of the text. He advises hypertext designers to attend to points of 'arrival and departure', arguing that they 'must decide what readers need to know at either end of a hypermedia link in order to make use of what they find there'. Landow thinks that authors of hypertext materials confront three related problems: how to orient readers so that they can read efficiently and with pleasure; how to devise ways in which to inform readers where the links lead; and how to help readers who have just entered the document to feel comfortable there. For Landow, 'the most important aspect of hypertext design is *integration*, the assembly of parts into a meaningful whole' (Moulthrop, 1991b:152).

While acknowledging that the issues of navigation and orientation are critical in hypertext, Bernstein argues that the widespread belief that hypertext navigation is difficult appears to rest on three arguments, each of which can be countered because it is 'vulnerable' (1991:287). The first is that hypertext navigation necessarily adds a new and unfamiliar cognitive burden to the customary burdens associated with reading. The second is that hypertext readers issue a large number of navigational commands. And the third comprises anecdotes of how particular individuals found a specific hypertext document confusing. According to Bernstein, however, we need to know more of the cognitive processes inherent in reading before we can be certain that the navigation of a hypertext places any additional burdens on readers. Furthermore, the frequency with which users issue navigational commands could be simply an effect of specific interface designs. Finally, anecdotal evidence of disorienting experiences may signal problems with a particular document rather than with the medium itself. We have to distinguish 'intrinsic' navigational problems from those caused by inept writing and design.

Unlike databases, hypertexts are meant to be read, not searched; readers follow meaningful links or paths through the document instead of issuing queries. Unlike knowledge bases, hypertexts contain information meant to be understood by human readers, not machines. Effective hypertext writing depends therefore on the tension between regimentation and richness, between predictability and excitement: 'mild disorientation can excite readers, increasing their concentration, intensity, and engagement' (ibid.:295).

Metaphor and hypertext

If hypertext is the medium of the future, we will have to find metaphors that liberate our notions of 'text' from the conventions and confines of the printed page. What we need are metaphors that evoke what seem to be the essential features of hypertext technology, and at the same time have the power to extend our grasp of its potential. This process has already begun.

Unlike print, hypertext is a form in which the physical structure of a piece of writing need not match its logical structure. A metaphor currently used by hypertext designers is 'browsing', a surprisingly bookish term to describe the ability to roam through information in a manner which the structured path through a printed text does not condone. It may well trouble those accustomed to the visual cues provided by paper-based information.

'From a hypertextual computer screen, the reader can go almost anywhere, but also nowhere, if all the options are not readily apparent' (Shirk, 1991:185).

Another metaphor represents hypertext as a topological space to be navigated. Thus Bolter (1991:25) describes hypertextual writing as 'topographic', using the word in its original sense as a written description of a place (*topos*), such as an ancient geographer might give. To Bolter, hypertextual writing is both a visual and verbal description: 'It is not the writing of a place, but rather a writing *with* places' (ibid.). But he also points out that topographic writing is not computer-specific, since it is possible to write topographically in both print and manuscript cultures. 'Whenever we divide out text into unitary topics', he writes, 'and organise those units into a connected structure and whenever we conceive of this textual structure spatially as well as verbally, we are writing topographically' (ibid.). Although we do not need the computer to write topographically, hypertext accommodates this mode of writing most effectively.

Joyce (1995d:13) also uses a topological metaphor, speaking of the 'contours' in hypertext writing. For Joyce (1992:93), a contour is 'the space of inscription for a reader, the emerging surface of the constructive text as it is shaped by its reading'. The contour is a virtual representation of the reader's experience of the hypertext as it unfolds in time. It 'remains contingent as the reader moves back and forth through the text, revealing or realising new connections'. Each time the reader makes a link, the contour unfolds 'through a process of replacement' (Moulthrop, 1992:115).

When travel metaphors—'reading as voyage, excursus as excursion' (Harpold, 1991:126)—are used to describe movement through this topological space, hypertext is assimilated into 'the massive philosophical tradition that joins the language of travel, expedition and navigation to the practice of writing' (ibid.:127). That tradition incorporates such diverse forms as 'the Sophists' tours through the halls of memory, the Romantics' *promenades solitaires*, and Leopold Bloom's circuit between the headlines of the "Aeolus" episode of *Ulysses*' (ibid.). Much of Jacques Derrida's work—for example, the thread on travel, tourism and voyage that runs through the 'Envois' section of *The Post Card* (1980)—can be seen as a commentary on this tradition.

To some users, these travel metaphors suggest that facet of hypertext which Landow (1992a:11) calls 'infinitely recenterable'. They see the metaphor of 'navigation' as usefully evoking those active processes that readers must engage in while reading a hypertext: it highlights the dynamic role taken by the reader in deciding whether to land on or circumnavigate islands of information in a vast expanse of hypertext. Others think that the troping of hypertext as voyage 'is invariably used in a cautionary, negative sense' (Harpold, 1991:126). To Harpold, navigation 'presumes displacement, separation and loss, departures and farewells. New ports of call, perhaps, but also old ones that are visited less frequently than before' (ibid.:127).

Probably the most popular metaphor for hypertext is that of the web, which acknowledges the myriad of associative connections between written and pictorial texts. Another common metaphor is that of the maze or labyrinth. 'A labyrinth', writes John Barth (1967:34), 'is a place in which, ideally, all the possibilities of choice . . . are embodied, and . . . must be exhausted before one reaches the heart'. At its heart resides the Minotaur, who offers us two possibilities: 'defeat and death, or victory and freedom'. Like 'navigation', the metaphor of the labyrinth provokes contrasting responses in hypertext users. If it is perceived as an intricate structure of intercommunicating pathways—through which it is difficult but not impossible to find one's way without a clue, then the metaphor may evince hopes of conquest and encourage the belief that to find one's way through a maze involves challenges, adventure and excitement. If, however, the labyrinth is taken to be an obscure and even tortuous arrangement, then the metaphor may suggest that to find one's way through it is impossible, and can result only in confusion, frustration and ultimate defeat.

In Borges's (1970c) story, 'The Garden of Forking Paths', the idea of the labyrinth is described so intriguingly as to arouse desire in both the narrator and the reader to know more about it. The story—which appears in a collection called *Labyrinths*—is a mystery involving a quest for a labyrinth, and is based on 'the curious legend that Ts'ui Pen had planned to create a labyrinth which would be strictly infinite' (ibid.:50). Although no one finds the labyrinth, the narrator of the story remains enthralled:

> Beneath English trees I meditated on that lost maze: I imagined it inviolate and perfect at the secret crest of a mountain; I imagined it erased by rice fields or beneath the water; I imagined it infinite, no longer composed of octagonal kiosks and returning paths, but of rivers and provinces and kingdoms . . . I thought of a labyrinth of labyrinths, of one sinuous spreading labyrinth that would encompass the past and the future and in some way involve the stars. (ibid.:48)

Even though the quest for the labyrinth in Borges's story failed, the search was not without meaning. Borges's evocative conception of the infinite labyrinth serves as a potent metaphor for our pursuit of knowledge in the vastness of hypertext: an active, exhilarating, if somewhat frustrating endeavour. Borges's labyrinth embodies the understanding of hypertext central to this book.

3

Reconceiving Textuality

Hypertext and literary theory

The use of hypertext invites us to rethink and to redefine textuality in ways which correspond with ideas central to contemporary literary theory, as presaged in books and essays published around the time when hypertext technology was being conceived but had not yet become available. Hypertext theorists—Landow (1992a), Bolter (1991), Moulthrop (1991c), Tuman (1992a), Lanham (1993) and Johnson-Eilola (1994)—all agree that the connections between hypertext technology and recent literary theory are not only significant but also express a relationship more complex than that of mere contingency, in so far as 'both grow out of dissatisfaction with the related phenomena of the printed book and hierarchical thought' (Landow, 1994:1). When identifying the often dramatic nature of these connections, theorists articulate their insights in complementary, but subtly contrasting, ways.

Landow believes that the parallels between hypertext and what he calls 'critical theory' are so strong that the two fields have 'converged'. Although not necessarily aware of one another's writings, both hypertext theorists and critical theorists agree that 'we must abandon conceptual systems founded upon ideas of centre, margin, hierarchy, and linearity and replace them with ones of multilinearity, nodes, links, and networks' (Landow, 1992a:2). Landow begins his book on *Hypertext* (1992)—subtitled

The Convergence of Contemporary Critical Theory and Technology
—with the observation that 'when designers of computer software examine the pages of *Glas* or *Of Grammatology,* they
encounter a digitalised, hypertextual Derrida; and when literary
theorists examine *Literary Machines,* they encounter a decon-
structionist or poststructuralist Nelson' (ibid.). According to
Landow,

> critical theory promises to theorise hypertext and hypertext
> promises to embody and thereby test aspects of theory,
> particularly those concerning textuality, narrative, and the roles
> or functions of reader and writer. Using hypertext, critical
> theorists will have, or now already have, a new laboratory, in
> addition to the conventional library of printed texts, in which
> to test their ideas. Most important, perhaps, an experience of
> reading hypertext or reading with hypertext greatly clarifies
> many of the most significant ideas of critical theory. (ibid.:3)

Hypertext has so much in common with such matters as Derrida's
emphasis on 'decentring' and Barthes' conception of the 'writerly'
text (which co-opts its readers as co-creators) that it constitutes
'an almost embarrassingly literal embodiment' (Landow and
Delany, 1991:6) of both concepts.

Presenting his case a little differently, Bolter concludes that
'hypertext is a vindication of postmodern literary theory' (1992a:
24). He points out that for 'the past two decades, postmodern
theorists from reader-response critics to deconstructionists have
been talking about text in terms that are strikingly appropriate to
hypertext':

> When Wolfgang Iser and Stanley Fish argue that the reader
> constitutes the text in the act of reading, they are describing
> hypertext. When the deconstructionists emphasise that a text is
> unlimited, that it expands to include its own interpretations—
> they are describing a hypertext, which grows with the addition
> of new links and elements. When Roland Barthes draws his
> famous distinction between the work and the text, he is giving
> a perfect characterisation of the differences between writing in
> a printed book and writing by computer. (ibid.)

Bolter finds it 'uncanny' that many of those postmodern pronouncements which scandalised print-bound readers seem no more than descriptive of the properties of computer-generated hypertext:

> The irony is that postmodern theorists have been doing this without knowing it. Their methods have always been directed in the first instance to printed or written texts, not to electronic writing; their aim has been to upset our complacent notions of conventional printed literature. And their work has been controversial, to many shocking, when applied to literature in print. But when we read the same work with the computer in mind, it becomes indisputable, obvious, indeed (as mathematicians like to say) almost trivially true. (ibid.)

In many ways, suggests Bolter, hypertext confirms what deconstructionists and other contemporary theorists have been saying about the instability of the text and the decreasing authority of the author. 'What is unnatural in print becomes natural in the electronic medium and will soon no longer need saying at all, because it can be shown' (Bolter, 1991:143).

In the opinion of Moulthrop, those 'challenges to our reception and conception of text' which have accompanied the arrival of hypertext 'are really accelerated repetitions of movements that have been running through the literary world since the fifties and sixties' (1991c:255). He points out that figures like Barthes, Derrida, Michel Foucault and Fish envisioned a user-centred literature. They have

> compellingly criticised the duplicitous authority of writing, its claim to evoke a presence in which it does not participate, its attempt to articulate a discourse which in fact it delimits or betrays. By way of an alternative, they ask us to consider language not as a hierarchy but as a network of relationships: the model for written discourse is no longer a linear chain of reference but a recursive, allusive web of correspondences. (ibid.)

The printed page and the bound volume, argues Moulthrop, are the 'wrong media for this new perspective on writing' (ibid.).

He suggests that what Lanham (1989:265) calls a 'digital revolution' —signalled most significantly by the arrival of fully electronic text—is

> a first step away from what Gilles Deleuze and Felix Guattari (1987:8) call 'arborescent culture', in which the controlling model of discourse is the hierarchical or genealogical tree, where branches and offshoots are subordinated to a single taproot. The coming changes in textuality allow us to create a different kind of linguistic structure, one that corresponds more closely to Deleuze and Guattari's 'rhizome', an organic growth that is all adventitious middle, not a deterministic chain of beginnings and ends. (Moulthrop, 1991c:254)

Although the cultural establishment has been able to restrict such destabilising ideas by confining them to the 'hothouse of literary theory . . . the current direction of technological development suggests that this resistance is no longer feasible' (ibid.:254–5).

Tuman (1992a:62) suggests that certain texts by post-structuralist theorists—such as 'From Work to Text' by Roland Barthes (1979)—can be read as prefatory to 'a new computer-based, hypertextual notion of text'. In the world of on-line literacy, 'the object of exchange will not be the individual text but literature itself, defined as the full extension of the connective and storage capacity inherent in hypertext' (ibid.:61). In other words, it will be Nelson's docuverse, accessible to us all through whatever machine we use. 'In a world defined by a single literature rather than a multitude of texts, reading becomes essentially a means of finding one's way—one moves, not ever deeper into a single text in quest of some world-altering hermeneutic understanding, but playfully between texts, from side to side as it were' (ibid.:62). Hypertext thus supports

> the wider contemporary movement away from a serious, introspective, relentlessly psychological (and often Germanic) hermeneutic tradition of interpretation—one often associated with modernism, despite its unmistakable nineteenth-century romantic origins—and toward a decidedly more ludic (and often Gaelic) postmodern concern with defining reading, and cultural criticism generally, as the play of signs.

This shift, argues Tuman, alters 'our primary understanding of the central terms of *text, author,* and *reader*' (ibid.).

Like Landow (1992a), Lanham also identifies an 'extraordinary convergence between technological and theoretical pressures' (1989:279), which he would later call a 'curious case of cultural convergence' (1993:132). Lanham believes that western culture, 'for which "the Great Books" has come to be a convenient shorthand phrase, is not threatened by the world of electronic text, but immensely strengthened and invigorated' (ibid.). He argues that 'digitising the arts requires a new criticism of them', and that 'we have it already in the postmodern aesthetic. The fit is so close that one might call the personal computer the ultimate postmodern wórk of art' (ibid.:108). He goes on to suggest that

> if we look at the world of electronic text from the perspective of what we have come to call literary theory, an electronic world comes not as a technological *vis a tergo* but as a fulfil-ment. The conceptual, even the metaphysical, world that digital text creates—dynamic rather than static, bi-stable rather than mono-stable, open-ended, rather than self-contained, partici-patory rather than authorial, based as much on image and sound as on word—*is* the world of postmodern thought, the world which . . . began with Italian Futurism and Dada and is now the focus of theoretical discussions in disciplines all across the human sciences. In introducing our students to electronic text in the practical world of work, it turns out that we also introduce them to the critical issues of our intellectual life. (ibid.:132–3)

The electronic word, as pixeled upon a personal computer screen, replicates the logics of postmodern thought by 'literalis[ing] them in a truly *uncanny* way' (Lanham, 1989:287).

To Johnson-Eilola technologies such as hypertext give theorists and teachers

> the opportunity to remap their conceptions of literacy, to reconsider the complex, interdependent nature of the ties between technology, society, and the individual in the acts of writing, reading and thinking. Adding the concept of hypertext to theory does not replace other definitions or conceptions of

writing and reading; it opens those definitions up to debate and change. (1994:204)

He believes that hypertext helps us to 'revise theories of reading, writing, and literacy in key ways by making various traits of these theories visible' (ibid.:203). He explains that the text can be 'deconstructed [not only] in the reader's mind or in a secondary, parasitical text, but also visibly on the computer screen' (1993:382).

He also argues that 'the parallels between hypertext and contemporary writing and reading theory suggest that the radical alterations in literary theory that hypertext seems to necessitate are not completely new prophecies: these changes were merely predicted in relation to print text' (1994:196). More than any previous text technology, hypertext 'encourages both writers and readers —roles we might now provisionally combine under the label of hypertext "writer/readers"—to confront and work consciously and concretely with deconstruction, intertextuality, the decentring of the author, and the reader's complicity with the construction of the text' (1993:382). He stresses, however, the tentative nature of any claims that can be made for the technology, and is careful to talk only of 'the potential' of hypertext to make visible the goals and processes of current theory (1994:217).

'Convergence', 'embodiment' and 'literalisation' are metaphors commonly used by these writers to characterise the relationship between hypertext and contemporary literary theory. Others are 'dramatisation', 'instantiation' and 'materialisation'. While each metaphor conveys a slightly different representation of how the two fields interconnect, collectively they convey a strong impression of the significance and complexities of such interrelations and commonalities. One interesting difference between the writings of literary and hypertext theorists, however, is tone. Most writings on theory—with the notable exception of Derrida's—are models of scholarly solemnity; by contrast, writings on hypertext are most often celebratory. 'Whereas terms like *death, vanish, loss*, and expressions of depletion and impoverishment colour critical theory', notes Landow (1992a:87), 'the vocabulary of freedom, energy, and empowerment marks writings on hypertextuality'. He attributes this difference in tone to the fact that critical theorists work with the limitations of print; hypertext advocates, on the other hand, revel in possibility, 'excited by the future of textuality, knowledge, and writing' (ibid.). Yet if writers on hypertext are to appear both credible and persuasive to a sceptical readership, they need to adopt a more moderate tone, especially when considering claims that the new technology has the potential to alter social structures. The combination of a politically partisan vocabulary with an often zealous tone tends to reveal more about the writers' own ideologies and agendas than about the powers of hypertext.

Abandoning the linear

Although the printed book allows the writer to suggest—by such rhetorical devices as ambiguity and irony—that there are alternative paths through the same work, in most books one path dominates: 'the one defined by reading line by line, from first page to last' (Bolter, 1991:108). This constitutes the 'canonical order' (ibid.) of a printed book; and even though it can be read out of order, the printed sequence is suggestive and controlling. This trend toward linear and hierarchical structures is evident even in manuscript culture, but was reinforced by the printing press. Subsequently, the oral character of text—its digressions and dependence on the visual—declined. Today such major forms of non-fiction as the essay, the treatise, and the report 'are expected to be hierarchical in organisation as they are linear in presentation' (ibid.:112). This is the model for scholarly and scientific as well as for business and technical writing.

The computer, however, calls into question the belief that the written text must take the form of a linear progression. Why do linearity and exclusiveness persist as stylistic ideals when the new electronic writing space made available by hypertext allows a writer to entertain and present several lines of thought at once? This question was posed even before the arrival of microcomputers by Roland Barthes, who broke down linear forms in producing texts characterised by fragmentation and interruption. But although Barthes rejected linear modes of argument, he retained the physical form of the printed book. Derrida's *Glas* (1976a), however, disrupts our notion of what a book should look like. Each of its pages is divided into two columns: the left offers passages from glossed extracts from the philosophical writings of Hegel, while the right is a commentary on the French novelist, Jean Genet. Paragraphs set in and around other paragraphs—presented in variable sizes and styles of type—'give the page an almost medieval appearance' (Bolter, 1991:116). Although no linear argument spans the columns, the reader's eyes are drawn across, down and around the page in search of the apparent visual and verbal connections between Hegel and Genet texts. 'In *Glas* Derrida lays down a textual space and challenges his reader to find a path through it', Bolter observes. 'Whatever else he is doing, Derrida is certainly writing topographically, as if for a medium as fluid as the electronic' (ibid.).

In *Of Grammatology* (1976b), Derrida argues that non-linear writing has been suppressed though never eradicated by linear writing. Believing that modern experience could not be recorded adequately in linear forms, Derrida concluded that we would begin to write in new ways. In this, argues Bolter, 'Derrida was prescient, but he could not know that electronic writing would be the new writing to which he alluded' (1991:116). Derrida (1976b:86) thought that 'the end of linear writing is indeed the end of the book'. Instead, suggests Bolter, 'the new electronic medium [in the form of hypertext] redefines the book in a way that incorporates both linear and nonlinear form' (1991:116).

Although hypertext allows readers to choose from multiple paths through a body of text, it is not wholly devoid of linearity. After all, 'intelligibility in language demands some kind of sequence; or in the case of hypertext, some level of absolute sequentiality' (Moulthrop, 1992:114). No one would seriously propose a hypertext system in which 'variability or randomness obtain at the character level' (ibid.). All hypertexts therefore retain some degree of sequentiality, 'some minimal dimension or granularity—word, phrase, paragraph, chapter—at which the reader is momentarily disengaged from the variability of the text' (ibid.:115). Landow (1992a:4) follows Barthes in calling these chunks of text 'lexias'. Lexias are units of local stability in the general flux of the hypertext. The local stability of the lexia arouses expectations of coherence and internal consistency, which are the familiar hallmarks of print. But the operation of the link overturns such expectations, throwing the reader into unfamiliar discursive territory and invalidating structures like causality or necessity. 'No wonder', Moulthrop points out, 'hypertext seems so problematic to researchers interested in coherence and unity: the experience of hypertextual reading is fundamentally dissonant' (1992:115).

Text as network

The notion of the text as network is familiar from the literary theory of Roland Barthes, who in *S/Z* (1974) distinguishes the 'work' (the object of traditional literary study), from the 'text', the new field of discourse which he sought to develop. The 'work' is a defined body of writing, bound in a volume marked with an author's name, and sanctioned and validated by tradition. It is the material object we are accustomed to: it can be held in the hand, purchased in bookshops and borrowed from libraries. Against this 'classic' idea of the literary 'work' Barthes proposes the rival notion of the 'text', which is a web of language that links the 'work' to other discourses, some of which are creative, others critical, and some not even literary. The Barthesian 'text' is not so much a new object as a new attitude. It views reading as the site of an activity, a place where the experience of reading suddenly branches out in many directions in order to establish links with an ever expanding world of meanings.

Barthes' theory of the text is 'a precursor of the hypertextual notion of all writing as part of a single universal literature' (Tuman, 1992a:63). In Barthes' own words:

> The Text is not coexistence of meanings but passage, traversal; thus it answers not to an interpretation, liberal though it may be, but to an explosion, a dissemination. The Text's plurality does not depend on the ambiguity of its contents, but rather on what could be called the *stereographic plurality* of the signifiers that weave it. (1979:76)

For Barthes, the defining metaphor of the work (like writing before hypertext) is that of an organism, something 'that grows by vital expansion, by "development"'. 'The Text', Barthes adds, 'is read without the father's signature. The metaphor that describes the Text is also distinct from that describing the work . . . The Text's metaphor is that of the *network*' (ibid.:78).

Barthes describes the text as 'a system without end or centre' (ibid.:76), which is an equally appropriate description of hypertext. 'A text without end or centre is in our normal usage not a text at all but literature itself' (Tuman, 1992a:63). Barthes reverses this in *S/Z*, when he says that 'literature itself is never anything but a single text: the one text is not an (inductive) access to a Model, but entrance into a network with a thousand entrances' (1974:12). To read this text, in Barthes' words, is to take 'aim', not at an object, but at a 'perspective (of fragments, of voices from other texts, other codes)' (ibid.). As Tuman explains, 'the object of our reading, this "perspective", is obviously a text only in some new and, in terms of print technology, seemingly paradoxical sense' (1992a:63). The object of our reading is a text in the form of hypertext.

Derrida similarly conceives of text as constituted by discrete units of reading which are separate yet bound together. According to Landow (1992a:8), in *Glas* Derrida 'acknowledges . . . that a new, freer, richer form of text, one truer to our potential experience, perhaps to our actual if unrecognised experience, depends upon discrete reading units'. In the case of every mark spoken and written, Derrida sees a possibility of disengagement. Every sign, linguistic or non-linguistic, can be cited, put between quotation marks. The implication of such separability is that here, as in hypertext, every reading unit 'can break with every given context, engendering an infinity of new contexts in a manner which is illimitable' (Derrida, 1977:185). The resulting 'assemblage' —as Derrida calls it—has 'the structure of an interlacing, a weaving, or a web, which would allow the different threads and different lines of sense or force to separate again, as well as being ready to bind others together' (1973:131).

Our subsequent experience of hypertext, argues Landow (1992a:9), shows how Derrida 'gropes toward a new kind of text' in his early writings, but is constrained by the medium of print in which he is compelled to express his ideas: he can 'present it only in terms of the devices—here those of punctuation—associated with a particular kind of writing'. That new type of text which eludes Derrida materialises later in Nelson's conception of hypertext as docuverse, or what Landow calls the '*metatext*' (ibid.). Both Derrida and Nelson describe a montage-like notion of textuality, a web of ideas that separate or bind different lines of meaning by reciprocally informing one another. Landow depicts Derrida's conception of text as 'an instinctive theorising of hypertext' (ibid.). *Glas* and hypertext both work deliberately to make obsolete the linear convention of printed books and to replace them with the more complex, networked, multi-linear and multimedia hypertext.

Like Barthes and Derrida, Foucault conceives of text in terms of networks and links. He argues that because 'the frontiers of a book are never clearcut', the text is 'caught up in a system of references to other books, other texts, other sentences: it is a node within a network . . . [a] network of references' (1976:23). Foucault models his own project—'the archaeological analysis of knowledge itself' (ibid.)—as a network capable of linking together 'a wide range of often contradictory taxonomies, observations, interpretations, categories, and rules of observation' (Landow, 1992a:25).

When considering the connections between contemporary critical theory and the designs of hypertext, Landow notes their mutual emphasis on 'the model or paradigm of the network' (ibid.:23), and identifies four different usages of the term in descriptions of hypertext systems. First, when print is transferred to hypertext it takes the form of blocks or lexias joined by a 'network' of links and paths. Second, a set of lexias in hypertext, whether collected by one author or many, takes the form of a 'network' or a web. Third, a 'network' describes a system that connects computers so that information can be shared: these can be Local Area Networks (LANs), which join machines within an institution, or Wide Area Networks (WANs), which link institutions globally. But when 'network' is related to hypertext, Landow believes, it 'comes close to matching the use of the term in critical theory':

Network in this fullest sense refers to the entirety of all those terms for which there is no term and for which other terms stand until something better comes along, or until one of them gathers fuller meanings and fuller acceptance to itself: 'literature', 'infoworld', 'docuverse', in fact 'all writing' in the alphanumeric as well as Derridean senses. The future wide area networks necessary for large scale, interinstitutional and intersite hypertext systems will instantiate and reify the current information worlds, including that of literature. To gain access to information, in other words, will require access to some portion of the network. To publish in a hypertextual world requires gaining access, however limited, to the network. (ibid.:24)

Pagels explains why the notion of a network appeals to those who are wary of hierarchical or linear models: 'a network has no "top" or "bottom". Rather it has a plurality of connections that increase the possible interactions between the components of the network. There is no central executive authority that oversees the system' (1989:50). This may explain why certain hypertext theorists are so enamoured (perhaps naively) of what they believe this technology promises: namely, a democratisation of access to information, knowledge and the academy.

The open text

Hypertext creates 'an open, open-bordered text, a text that cannot shut out other texts' (Landow, 1992a:61). This open text 'embodies' (ibid.) the Derridean text, which blurs 'all those boundaries that form the running border of what used to be called a text, of what we once thought this word could identify, i.e., the supposed end and beginning of a work, the unity of a corpus, the title, the margins, the signatures, the referential realm outside the frame, and so forth' (Derrida, 1979:83). Indeed, Derrida's characterisation of a text sounds very much like 'text in the electronic writing space' (Bolter, 1991:162). And yet, when Derrida speaks of 'marginality', or of the text as extending beyond its borders, 'he is in fact appealing to the earlier technologies of writing, to medieval codices and printed books' (ibid.). Like other contemporary theorists, Derrida sets out to reverse a literary hierarchy while 'assuming the technology of printing (or sometimes handwriting) that generates or enforces that hierarchy' (ibid.:163). If the margins that concern Derrida are 'the borders of the printed or written page' (ibid.), what would he say about electronic texts? In hypertext the only margins and boundaries are 'the ultimate limitations of the machine' (ibid.). Hypertext supports a network in which all the constituent elements have such equal status that 'to be at the margin is itself only provisional' (ibid.). In hypertext,

> the author can extend and ramify this textual network limited only by the available memory. The reader can follow paths through the space in any direction, limited only by constraints established by the author. No path through the space need be stigmatised as marginal. (ibid.)

In hypertext systems, 'links within and without a text—intratextual and intertextual connections between points of text (including images)—become equivalent, thus bringing texts closer together and blurring the boundaries among them' (Landow, 1992a:61).

Because hypertext redefines what constitutes the borders of a text, notions of 'inside' and 'outside' no longer apply, and ideas about textuality subsequently change. The material requirements of printing and publishing created 'containers' (Delany and Landow, 1993:15) which standardised the size and format of the printed text. Digitised and networked, however, texts 'smash the containers' (ibid.). In a hypertext environment, a scholarly article is linked to all the materials it cites, and is therefore perceived as part of a larger system in which the totality counts more than the individual document. The post-structuralist concept of the 'open text' is realised literally by the computer's capacity 'to manipulate, to disperse, and to recombine the elements of digital texts' (ibid.).

According to Moulthrop (1991b), when Barthes (1979) differentiated 'the work' from 'the text', he was distinguishing between writing as volume and writing as virtual space. Barthes' theory of discourse portrays the text not as a discrete territory but as a polyvocal and intertextual *'social* space' (Barthes, 1979:81) of writing. 'Texts produced in this communal space would be heterogeneous networks where the language of the author would be joined to those of readers, commentators, critics, biographers, and so on' (Moulthrop, 1991b:152). But such texts cannot exist in print—'books are not and never can be open discursive networks' (ibid.). No matter how strenuously we attempt to deconstruct it, print continues to impose closure and coherence on the space available for writing. 'The book is always in essence a *territorial* form of writing, an attempt to fix boundaries for thought or expression' (ibid.:153).

Because hypertext evokes integration rather than self-containment, it allows us to move beyond our notion of the book and to explore a truly textual space. Hypertext exploits 'the computer's capacity to designate any unit of text as a new element in an expanding vocabulary of signs' (Bolter, 1991:60). As multiple discursive sequences, hypertexts are not unitary documents but 'docuverses'. Each offers the 'possibility of a "deterritorialised" writing, a conception based not on circuitries of arrival/departure or "exploration" but instead on pure extension' (Moulthrop, 1991b:154). This is logically the next step after Balestri's 'softcopy': it corresponds with Joyce's notion of 'constructive' hypertext, and is the linguistic realisation of Deleuze and Guattari's 'rhizomatic' form. 'The text-as-hypertext is a field of language whose divisions and boundaries are always at issue, a discourse in which all forms of authority are provisional and contingent' (Moulthrop, 1991c: 257). Barthes (1974:6) contends that 'as nothing exists outside the text, there is never a *whole* of the text'. Hypertext begins 'to undo the book's conceptual sense of closure and authority' (Johnson-Eilola, 1992:111). Relatively open—both conceptually and physically—the boundaries of a hypertext are continually shifting as new texts and new links are added. This is why 'the guiding metaphor for hypertext is not the bound and complete text but the networked intertext' (ibid.), the truly open text.

Dispersal of the text

Whereas spatial fixity characterises the book text, electronic text is continuously various. Because no one state or version is ever final, an electronic text is relatively dynamic in permitting correction, updating and modification. Even at the level of simple word processing it is not constrained by the fixity that characterises print. But once the linking which is integral to hypertext has been introduced into the electronic text, fixity is abandoned altogether.

Hypertext introduces randomness into text by decreasing the possibility of control over its edges and borders. The text thus appears to fragment and to atomise into its constituent parts or blocks. Each takes on a life of its own as it becomes more self-contained and less dependent on what precedes or follows it in a linear succession. Once a hypertextual node has loosened its bonds with other nodes in the same work, it may associate itself immediately with text created by other authors, thereby dissolving notions of the 'intellectual separation of one text from others' (Landow, 1992a:53). Hypertext is thus 'promiscuous (in the root sense of "seeking relations")' text (Moulthrop and Kaplan, 1994:227).

Another effect of linking is to disperse 'the' text into other texts:

> As an individual lexia loses its physical and intellectual separation from others when linked electronically to them, it finds itself dispersed into them. The necessary contextuality and intertextuality produced by situating individual reading units within a network of easily navigable pathways weaves texts, including those by different authors and those in nonverbal media, tightly together. One effect of this process is to weaken and perhaps destroy any sense of textual uniqueness. (Landow, 1992a:53)

While such notions are hardly novel to literary theory, hypertext embodies 'a principle that had seemed particularly abstract and difficult when read from the vantage point of print' (ibid.). Moreover, another further fundamental variation enables it to disperse or atomise text, namely its provision of electronic links that permit readers to take different paths through blocks of text. This capacity to avoid unilinearity has important effects on conceptions of textuality and rhetorical structures.

In *S/Z*, Barthes explains his procedure for fragmenting the text of Balzac's *Sarrasine*:

We shall therefore star the text, separating, in the manner of a minor earthquake, the blocks of signification of which reading grasps only the smooth surface, imperceptibly soldered by the movement of sentences, the flowing discourse of narration, the 'naturalness' of ordinary language. The tutor signifier will be cut up into a series of brief, contiguous fragments, which we shall call *lexias*, since they are units of reading. (1974:13)

The process Barthes describes moves seemingly in the direction of hypertextuality, and disturbs both Balzac's text and what is commonly understood to be the reading experience. Passages of text that had followed one another in an apparently seamless and inevitably linear progression 'now fracture, break apart, [and] assume more individual identities' (Landow, 1992a:53). Furthermore, although the linear rhetoric of 'beginning, middle and end'—so well suited to a print medium—may continue to appear within blocks of text, it can no longer be used to structure discourse. What had come to be perceived as a 'natural' sequentiality is disrupted in an electronic medium that encourages readers to choose different paths.

Although these developments in literary theory and hypertext signify major changes to our notion of textuality, they do not constitute a forsaking of the 'natural', because the book itself is anything but a natural structure. It 'took all of 4,000 years to bring about' (McArthur, 1986:69). Whereas ancient and medieval scholastics 'conventionalise[d] the themes, plot and shapes of books' (ibid.), their conventions of book structure 'changed fundamentally with the advent of the printing press, which encouraged alphabetic ordering' (Landow, 1992a:57).

If 'the movement from manuscript to print and then to hypertext appears one of increasing fragmentation' (ibid.), we should note that in the case of a hypertext document 'fragmentation' does not imply the kind of entropy that it does in the world of print. 'Capacities such as full-text searching, automatic linking, agents, and conceptual filtering potentially have the power to retain the benefits of hypertextuality while insulating the reader from the ill effects of abandoning linearity' (ibid.).

Intertextuality

There are strong connections between hypertext and contemporary notions of intertextuality as described by Jonathan Culler:

> literary works are to be considered not as autonomous entities, 'organic wholes', but as intertextual constructs: sequences which have meaning in relation to other texts which they take up, cite, parody, refute, or generally transform. A text can be read only in relation to other texts, and it is made possible by the codes which animate the discursive space of a culture. (1981:38)

The printed book encourages us to think of the text as an organic whole, a unit of meaning independent of all other texts. Hypertext, however, gives us a unique opportunity to visualise intertextuality:

> Stressing connections rather than textual independence, the electronic space rewrites the possibilities of reference and allusion. Not only can one passage in an electronic text refer to another, but the text can bend so that any two passages touch, displaying themselves contiguously to the reader. Not only can one text allude to another, but the one text can penetrate the other and become a visual intertext before the reader's eyes. (Bolter, 1991:163–4)

The space of hypertext is 'a fundamentally intertextual system' (Landow 1992a:10) with the capacity to explore intertextuality in ways that page-bound text in books cannot match. Although print can be made to display intertextuality, it does not encourage it, because the book's existence as a bound object serves to separate its constituent pages from those in other books. The ease with which a conventionally parenthetical citation can become a hypertext link to a completely different text 'promotes an intertextual conceptual space' (Johnson-Eilola, 1992:112).

Using hypertext, the reader can make explicit though not necessarily intrusive those linked materials that are perceived as surrounding a text. Scholarly articles and books exemplify *explicit* hypertextuality in non-electronic form by using symbols that denote the presence of footnotes, whose purpose is to signal the existence and location of subsidiary texts to the main text. Literary works are often *implicitly* hypertextual, as in the concluding section of Edith Wharton's *Age of Innocence* (1920). When describing Newland Archer's visit to Paris with his son Dallas, Wharton 'alludes' to many other 'texts'—including images of Paris, the city's boulevards, buildings and artworks—that suffuse and inform Archer's thoughts, and relate to other passages in the novel. A hypertext presentation of the novel would link this section not only to such visual materials mentioned but also to other works by Wharton, biographies, critical commentaries, Scorsese's film, textual variants, and so on.

Hypertext can be used, therefore, to electronically link all the allusions and references in a text, both external ('intertextuality') and internal ('intratextuality'). The work done by the Intermedia Unit at Brown University by Landow (1992a:37) and his collaborators favours this approach. According to Landow, Tennyson's poem *In Memoriam* 'anticipates' hypertext: its 'proto-hypertextuality' results from its being an 'antilinear, multisequential poem' (ibid.:38). The *In Memoriam* web attempts to map the non-linear organisation of the poem by linking the relevant sections. Using Intermedia's capacity to create bi-directional links—and to join an indefinite number of them to any passage of text—readers can move through the poem along many different axes. The web's link paths enable us to organise Tennyson's poem in terms of its intratextual leitmotifs. In the manner of Barthes' (1974) treatment of *Sarrasine* in *S/Z*, hypertext users may impose their own divisions upon a work. Moulthrop, for example, has created a hypertext called Forking Paths, which activates much of the potential for variation in Borges's short story 'The Garden of Forking Paths'. A theoretically illuminating project would be to create hypertexts of *Glas* and *S/Z*.

Multiple beginnings and endings

Hypertext reconfigures the text in another important way by redefining 'beginnings and endings' (Landow, 1992a:57) in non-linear terms as multiple rather than singular events. A 'beginning', Edward Said argues, with printed texts in mind, 'is designated in order to indicate, clarify, or define a *later* time, place, or action. In short, the designation of a beginning generally involves also the designation of a consequent intention' (1985:5). Thus even an atomised text can make a beginning when the link site, or point of departure, assumes the role of the beginning of a chain or path. As Said explains, 'we see that the beginning is the first point (in time, space, or action) of an accomplishment or process that has duration and meaning. *The beginning, then, is the first step in the intentional production of meaning*' (ibid.—original emphasis). Drawing on Said's work on origins and openings, Landow (1992a:58) argues that, unlike print, hypertext 'offers at least two different kinds of beginnings', one made up of the individual lexias, and the other constituted by the gathering of them into a whole text. In the case of hypertext, the beginning can never be more than the point at which one begins reading.

Hypertext also changes our conception of an ending. Different readers can choose not only to end the text at different points but also to add to and extend it. In hypertext there is no final version, and therefore no last word: a new idea or a reinterpretation is always possible. Bakhtin's (1984) conception of textuality antici-pates the kind of endings instantiated in hypertext. For Bakhtin, the whole is not a finished entity, because when it is realised it is open to change. Hypertext similarly undermines traditional notions of completion. If you were to put James Joyce's novel *Ulysses* (1922)—'conventionally considered complete' (Landow, 1992a:59)—into a hypertext format, it would immediately become 'incomplete'; so too would the *Encyclopaedia Britannica*. By privileging intertextuality, hypertext provides a large number of points to which other texts can attach themselves. Whereas the fixity of the book closes off such options, 'hypertext opens them up' (ibid.).

Decentring the text

Hypertext provides 'an infinitely re-centrable system whose provisional point of focus depends upon the reader' (Landow, 1992a:11). Its fundamental characteristic is that it is composed of bodies of linked texts that have no primary axis of organisation.

Although the absence of a centre can create problems for both readers and writers, it enables users to make their interests 'the de facto organising principle (or centre) for the investigation at the moment' (ibid.:12). Imagine, for example, a hypertext system devoted to scholarship on Patrick White. Any article on White could be linked to the whole range of other materials it cites. Received as part of a much larger system whose totality might count more than the individual document, such an article would appear to be woven more tightly into its context than its print-bound counterpart. In such a system, readers would traverse this web or network of texts so easily that they would be shifting the centre and hence the focus of their experience.

The differences between print and electronic text further our understanding of hypertext as decentred text. The printed page 'privileges exclusive and singular utterance' (Moulthrop, 1991c:260); for as Fish (1980) has pointed out, print fosters the illusion of an exhaustive, predetermined 'content'. Electronic text, on the other hand, 'lacks the single authority of a discursive centre' (Moulthrop, 1991c:260), because it presents itself

> as a field of linkage and associational play whose meaning depends upon its permutations. In reading as well as writing, electronic text thus erases the invidious distinction between process and product by allowing us to reformulate the text not as a limited artefact nor as a theoretical 'heteroglossia', but as a medium for the actual intersection of discourses. (ibid.)

Hypertext disrupts the convention of using footnotes or endnotes to incorporate information that is difficult or impossible to include in linear text. This convention rests on the distinction between a 'main' and a 'subsidiary' text. It establishes the dominance of authorial arguments over those to be found in the writings of scholarly allies and critics. In hypertext, however, textual annotation is a very different experience. Electronic linking immediately destroys that simple binary opposition of text and note on which the status relations that characterise the printed book are based. Hypertext makes it difficult to assign differential status to these two kinds of text. Consequently, in the electronic medium such textual hierarchies tend to collapse.

In *S/Z,* Barthes (1974) comments on the hierarchical relationships among portions of standard scholarly text as a political problem. Playfully creating his own version of complex footnote systems, he forces us to think about how we endow certain assemblages of words with power and value because they appear in a different format from the rest. In this respect, *S/Z* is a critique of the power relations between text and footnote. Barthes makes us see that although the devices designed to aid information retrieval—notes, prefaces, dedications, chapter titles, tables of contents and indexes—serve as a means of reader orientation, they have become so naturalised that we do not recognise them as conventions. In *S/Z,* he renders them visible by making us think about the textual politics of the footnote. For although the distinction between text and footnote establishes the primary importance of the 'main' text in its relation to other texts, such hierarchical relationships do not in fact exist.

Derrida also comments on the status relations that cut and divide texts, concentrating on oppositions not only between preface and main text but also between main text and other texts. He argues that the preface must be distinguished from the introduction because each has a different function and status. By proposing an entirely different analysis of such categorisation—in which the outlines of the preface and the 'main' text are blurred—Derrida (1976b) conceives of textuality in ways which resemble hypertextuality.

Hypertext problematises assumptions about the centrality of the text that is being read—'the linked text, the annotation, exists as the *other* text, and it leads to a conception (and experience) of text as Other' (Landow, 1992a:69). Because hypertext does not grant centrality to any node, what in print culture would be described as the 'marginal' comes to be as important as the 'central'. By dissolving centrality, hypertext models a society 'in which no one conversation, no one discipline or ideology, dominates or founds the others' (ibid.:70). As Landow argues, 'the basic experience of text, information, and control, which moves the boundary of power away from the author in the direction of the reader, models such a postmodern, antihierarchical medium of information, text, philosophy, and society' (ibid.).

Hypertext thus alters profoundly our notions of textuality. Hypertext is a plural text without a discursive centre, without margins, and produced by no single author. As an electronic text that is always changing and becoming, it is associative, cumulative, multi-linear and unstable.

4

Reconceiving Reading and Writing

Redefining the author

With the obvious exception of reference works and the like, books are written to be read in the order set out by the author. Books connote a sense of truth because they can be held, analysed, and quantified, and not so much read as consumed. Books are not, according to the Romantic view of them, the result of collaboration or negotiation, but instead are given by their authors so that they may be taken by their readers. 'They are also the model of all other cultural artefacts: paintings, plays, films, TV, buildings. In each case, the viewer, audience, inhabitant, whatever, is part of a normally non-negotiable pact with the painter, director, producer, architect, whoever, a pact which involves producing on the one part and consuming on the other' (Woolley, 1992:153). With the development of hypertext, this distinction between author and reader begins to lose its validity.

In the age of manuscripts, when scribes frequently altered what they copied, the distinction between authors and readers was not so significant. It was the invention of the printing press that strengthened the authority of the author. Because printing was a costly and laborious task, few readers had the opportunity to become published authors. Furthermore, because print had the effect of distancing authors from readers and making them less accessible, 'the author's words became harder to dismiss'

(Bolter, 1991:149). The printed copy also had more authority 'because of its visual simplicity, regularity, and reproducibility' (ibid.). Nowadays, successful authors are celebrities honoured both for their power to entertain and their presumed insight into the human condition. Moreover, through the technology of printing, the author (assisted by the editor and publisher) exercises absolute control over the text. Nothing can be done to it after publication.

The tradition of print literacy was thus organised in the service of a dominant author, a god-like figure who was normally male. It is his wishes, as expressed in the literary work, that readers (rendered more or less passive by this tradition) are supposed to respect. Hypertext and contemporary literary theory, however, 'reconfigure' (Landow, 1992a:71) the author in a number of important ways. First, in both hypertext and literary theory, the functions of reader and writer become deeply intertwined. This hypertextual 'convergence' of reader and writer infringes upon the power of the writer, 'removing some of it and granting it to the reader':

> One clear sign of such transference of authorial power appears
> in the reader's abilities to choose his or her way through the
> metatext, to annotate text written by others, and to create links
> between documents written by others. Hypertext . . . narrow[s]
> the phenomenological distance that separates individual docu-
> ments from one another in the world of print and manuscript.
> In reducing the autonomy of the text, hypertext reduces the
> autonomy of the author. (ibid.)

Hypertext gives to readers the power that had once been the prerogative of the author. In hypertext, consequently, the reader can become an active, independent and autonomous constructor of meaning.

In his two famous essays, 'From Work to Text' (1979) and 'The Death of the Author' (1993), Barthes writes about texts, authors and readers in ways that bear a remarkable resemblance to the realisation of these constructs in hypertext. At the end of 'From Work to Text', he describes how his open-ended notion of text will affect our traditional understanding of what it means to be an author. It will involve 'an attempt to abolish (or at least to lessen) the distance between writing and reading, not by intensifying the reader's projection into the work, but by linking the two together in a single signifying process' (Barthes, 1979:79). The text becomes the basis of collaboration.

In 'The Death of the Author', Barthes writes:

> We know now that a text is not a line of words releasing a single 'theological' meaning (the 'message' of the Author-God) but a multi-dimensional space in which a variety of writings, none of them original, blend and clash. The text is a tissue of quotations drawn from the innumerable centres of culture. (1993:116)

Here Barthes exposes the popular notion of 'a unified text as the expression of unified psyche' (Tuman, 1992a:64): in so doing, he provides 'a systematic literary analysis of the founding notion of online literacy, that computer-based writing is essentially author-less' (ibid.). If no text is ever more than an assemblage of fragments, its 'author' is merely the personage charged with collecting and arranging such material. 'Behind our traditional notion of the author as the unifying force responsible for creating the text . . . is not a human being but an historical construct' (ibid.).

Foucault once characterised authorship as 'the principle of thrift in the proliferation of meaning' (1979:159). Moulthrop argues that hypertext 'would revise this economy of language, empowering every user of the writing system either to reconstitute a given discourse by drawing his or her own links, or to expand the range of the document . . . by introducing new information' (1991c:257). In a limited but powerful way, electronic writing realises the type of text envisaged by Barthes as a '*social* space that leaves no language safe or untouched' (1979:81).

Both hypertext and literary theory break with traditional notions of authorship by configuring 'the author of the text *as a text*' (Landow, 1992a:72). As Barthes puts it: 'This "I" which approaches the text is already itself a plurality of other texts, of codes which are infinite' (1974:10). The 'author' of a text is therefore as intertextual a phenomenon as the text itself. If it is true that hypertext 'embodies many of the ideas and attitudes proposed by

Barthes, Derrida, Foucault, and others' (Landow, 1992a:73), then one of the most important 'involves treating the self of author and reader not simply as (print) text but as a hypertext' (ibid.). For all these theorists, the 'self' is not the source of meaning but merely the site of a decentred (or centreless) network of codes that, on another level, also serves as a node within another centreless network.

Hypertext accordingly changes our notions not only of textuality but also of authorship. The absence of textual autonomy and textual centredness impacts inevitably upon conceptions of authorship. In the unboundedness and openness of the new textuality made available by hypertext, the author is dispersed. Foucault notes the difficulty of defining this new notion of authorship:

> When undertaking the publication of Nietzsche's works . . . where should one stop? Surely everything must be published, but what is 'everything'? Everything that Nietzsche himself published, certainly. And what about the rough drafts for his works? Obviously. The plans for his aphorisms? Yes. The deleted passages and the notes at the bottom of the page? Yes. What if, within a workbook filled with aphorisms, one finds a reference, the notation of a meeting or of an address, or a laundry list: is it a work or not? (1979:143)

In response, Foucault replies: 'Why not? . . . it is not enough to declare that we should do without the writer (the author) and study the work itself. The word "work" and the unity that it designates are probably as problematic as the status of the author's individuality' (ibid.:144).

'There's more than one way to kill an author', suggests Landow (1992a:74). We can destroy the idea of sole authorship by denying autonomy to the text. We can also achieve the same end by decentring the text or by transforming it into a network. Finally, by removing the limits on textuality, we can permit it to expand until (as Foucault (1979:143) points out in his discussion of 'the disappearance—or death—of the author') Nietzsche the philosopher becomes equally the author of *Beyond Good and Evil* and laundry lists.

Said (1985) and Heim (1987) both claim that 'the erosion of the thinking subject' derives 'directly from electronic information technology' (Landow, 1992a:75). Said (1985:51) argues that 'the proliferation of information (and what is still more remarkable, a proliferation of the hardware for disseminating and preserving this information) has hopelessly diminished the role apparently played by the individual'. Heim (1987:220) believes that all electronic text entails loss of authorial power:

> Fragments, reused material, the trails and intricate pathways of 'hypertext', as Ted Nelson terms it, all these advance the disintegration of the centring voice of contemplative thought. The arbitrariness and availability of database searching decreases the felt sense of an authorial control over what is written.

As Heim implies, a database search permits an active reader to enter the text at other points than the one the author chose as the beginning. But because experienced readers have always behaved in this manner, 'the linear model of reading is often little more than a pious fiction' (Landow, 1992a:75).

The opportunity to participate in a network gives Heim (1987) more to worry about than celebrate because he connects loss of authorial control to loss of privacy. Even word processing redefines the notions of publishing, making public, and privacy. Because digital writing puts all users on a network, privacy 'becomes an increasingly fragile notion' (Heim, 1987:215). Whether or not one agrees with Heim, the issue he raises here has significant implications for an ethics of hypertext, and helps to account for the reactions of print authors 'accustomed to the fiction of the autonomous text' (Landow, 1992a:76).

Loss of authorial control in hypertext is considered also by Johnson-Eilola, who approaches the problem from the perspective of Donna Haraway's (1991) notion of the 'cyborg', in which the *cyb*ernetic melds with the *org*anic. In hypertext,

> control explicitly shifts away from the author, who begins to lose both the need and the opportunity for the great degree of control an author has in print because the hypertext writer's task is not to provide a narrow, fixed product but something

closer to a space for conversation with other texts, readers and writers. (Johnson-Eilola, 1993:382)

Hypertext, argues Johnson-Eilola, forces us to reconsider the writer's use of sequence in order to control the passage of readers through a text. In a hypertext environment, writers are never more than partially in control of their own processes, and therefore cannot fully control either the reader's path through the text or even the contents of the text itself. However, the control lost by the writer is not transferred directly to the readers of hypertext, for whom the computer—'as an active participant in the processes of writing/reading' (Johnson-Eilola, 1993:382)—becomes a way of structuring and navigating a text which they themselves cannot fully control.

The degree of control encountered in a hypertext results partly from the original author's manipulations and partly from the type of software and hardware used. The computer responds to a reader's movements through the text by managing shifts in structure across writing and reading time. When a reader makes navigational choices in each node of a hypertext like Izme Pass (Guyer and Petry, 1991), a series of if/then/else structures (hidden in each link) partially determines which node the reader will see next. These control structures vary the consequences of clicking on the same word in repeated readings of the same node, depending on what the reader has previously read. Identical navigational choices within a single node visited many times may result in very different outcomes for readers.

Johnson-Eilola (1993:387) is not arguing that the computer is 'intelligent', but that the interaction between human and machine-text requires us to view reading and writing hypertext as 'cyborg activities'. As Haraway (1991:66) explains: 'By the late twentieth century, our time, a mythic time, we are all chimeras, theorised and fabricated hybrids of machine and organism; in short, we are cyborgs'. To consider hypertext use as a 'cyborg activity' (Johnson-Eilola, 1993:384) is to understand that an ability to write elec-tronically is symptomatic of the fundamentally technological nature of our lives. Johnson-Eilola is not calling for 'technological "progress" or, conversely, Luddism', because we are 'neither moving forward to utopia nor backward to Eden; we are doing something else entirely' (ibid.). We must think critically, he urges, 'about what that something is', and 'consider our technological activities in a deeply social way' (ibid.).

A third form of reconfiguration of the author—'shared by theory and hypertext' (Landow, 1992a:76)—concerns the decentred self, which is 'an obvious corollary to the network paradigm' (ibid.). As Said points out, major contemporary theorists reject

> the human subject as grounding centre for human knowledge. Derrida, Foucault, and Deleuze . . . have spoken of contemporary knowledge (*savoir*) as decentred; Deleuze's formulation is that knowledge, insofar as it is intelligible, is apprehensible in terms of *nomadic centres*, provisional structures that are never permanent, always straying from one set of information to another. (1985:376)

According to Landow, these 'three contemporary thinkers advance a conceptualisation of thought best understood, like their views of text, in an electronic, virtual, hypertext environment' (1992a:77).

But we must also remember that although the unitary self occupies a privileged position in western thought, many thinkers have argued the contrary position.

> Divine or demonic possession, inspiration, humours, moods, dreams, the unconscious—all these devices that serve to explain how human beings act better, worse, or just different from their usual behaviour argue against the unitary conception of the self so central to moral, criminal, and copyright law. (ibid.)

We must realise 'that notions of the unitary author or self cannot authenticate the unity of a text' (ibid.).

A fourth change to traditional notions of authorship is that hypertext creates 'the virtual presence of all the authors who contribute to its materials' (ibid.:87). From the virtual texts encountered on the screen the reader constructs the virtual presence of those who contributed them. This is of course equally characteristic of those technologies of cultural memory which are based on writing and symbol systems: every record of an utterance conveys a sense of the one who made it. Hypertext differs from print technology, however, in amplifying the notion of virtual presence. 'The characteristic flexibility of this reader-centred information technology means, quite simply, that writers have a much greater presence in the system, as potential contributors and collaborative participants but also as readers who choose their own paths through the materials' (ibid.:88).

The virtual presence of other texts and other authors contributes to 'the radical reconception of authorship, authorial property, and collaboration associated with hypertext' (ibid.). Active readers necessarily collaborate with the author in producing a text by the choices they make. But collaboration occurs also when we compare one writer with the rest—'that is, the author who is writing *now* with the virtual presence of all writers "on the system" who wrote *then* but whose writings are still present' (ibid.).

In the Romantic tradition of authorship encountered in the humanities and social sciences, collaboration has not been encouraged. Hypertext changes our notion of authorship in the sense of abandoning the constrictions of page-bound technology. Having no 'authors' in the conventional sense, 'hypertext as a writing medium metamorphoses the author into an editor or developer. Hypermedia, like cinema and video or opera, is a team production' (ibid.:100).

Redefining the reader

A number of assumptions about reading underpin our print-based culture. One is that reading is a sequential, continuous and generally linear process. Readers are expected to begin at a clearly marked point, the appropriateness of which has been determined by the author: starting with the title, they continue in a sequential order through each paragraph or section, perhaps turning back from time to time to check a previous passage, or skipping ahead. Unless they lose interest in it, they read until they reach the end of the document. Another assumption is that readers predict what is going to happen next in a text, and do so on the basis of what they have read in this text and similar texts. A third assumption is that the text of a book is written or printed on paper and bound between covers: its stability as an object in physical space makes it seem equally stable as an object in mental space and encourages the desire to keep it that way forever. For all these reasons, sequentiality and textual stability are of paramount importance to readers in the print tradition.

Hypertext embodies a different set of assumptions about readers and reading. In conditions of electronic production, text 'is always mutable, always subject to inadvertent error and deliberate change, and it has to be coerced into standing still' (Slatin, 1990:872), usually by publishers who have the manuscript typeset and printed. In hypertext, 'reading is not assumed to be sequential and continuous' (Hager, 1993:104), but discontinuous, non-linear and associative, like the process of thinking. Hypertext enables text to be organised in new ways, driven by the reader's choice. Because the reader is now part of the text, the act of reading becomes correspondingly more conscious.

Reading in hypertext is a different experience from beginning at 'the beginning' and going through to 'the end'. Instead, readers begin at a point of their own choosing from a potentially large number of possibilities. They proceed by following a series of links that connect one document to another, and exit at any point they like, usually when they have had enough. It is thus common to encounter in discussions of hypertext metaphors which equate reading with the act of navigating or traversing large, open and sometimes poorly charted spaces.

What readers actually do when they move around in a hypertext web embodies many of the key concepts of reader-response criticism. For more than two decades, literary theorists examining the assumptions made about reading in print conditions have emphasised the role of the reader in the construction of textual meaning. Readers, they point out, respond actively to the words on the page, and it is their responses rather than authorial intentions which determine the text. 'The task of literary criticism, then, is not to examine the text in isolation, but rather to understand the text through its effect on the reader' (Bolter, 1991:156). Iser explains that the reader must participate in the literary act by realising the author's text:

> the literary work cannot be completely identical with the text, or with the realisation of the text, but in fact must lie halfway

between the two. The work is more than the text, for the text only takes on life when it is realised, and furthermore the realisation is by no means independent of the individual disposition of the reader . . . The convergence of the text and the reader brings the literary work into existence. (1980:50)

By Iser's account, the literary text is dynamic in nature and changes from reading to reading. Indeed, it is 'an arena in which the reader and author participate in a game of the imagination' (ibid.:51–2). The words on the page constitute the rules by which the game is played, but these boundaries still leave space for readers to use their imagination and fill in the gaps:

> one text is potentially capable of several different realisations, and no reading can ever exhaust the full potential, for each individual reader will fill in the gaps in his own way, thereby excluding the various other possibilities; as he reads, he will make his own decision as to how the gap is to be filled. (ibid.:55)

Iser also points out that the filling in of gaps is different with every act of reading, because 'the reading process always involves viewing the text through a perspective that is continually on the move' (ibid.:56). This observation is equally true of one's experience when reading hypertext, whose properties 'follow from the computer's ability to free the text from its embodiment as a set of bound, paper pages' (Bolter, 1991:157).

Fish also emphasises the paradox of the physical text:

> The objectivity of the text is an illusion, and moreover, a dangerous illusion, because it is so physically convincing. The illusion is one of self-sufficiency and completeness. A line of print or a page or a book is so obviously *there*—it can be handled, photographed, or put away—that it seems to be the sole repository of whatever value and meaning we associate with it. (1980:82)

As Fish acknowledges, although readers do not literally make the printed book, they shape the text as a structure of sounds, images and ideas in their own minds. 'This figurative text', notes Bolter, 'is for Fish, Iser, and others the only text worth studying'; but in the electronic medium,

> what was only figuratively true in the case of print, becomes literally true . . . The new medium reifies the metaphor of reader response, for the reader participates in the making of the text as a sequence of words. Even if the author had written all the words, the reader must call them up and determine the order of presentation by the choices made or the commands issued. There is no single univocal text apart from the reader; the author writes a set of potential texts, from which the reader chooses. (1991:158)

In a hypertext environment, readers acquire a new life by going on-line, which enables them to experience the seemingly separate acts of creating and reproducing as inextricably inter-woven events. Hypertext exemplifies Barthes' argument that the site where the multiplicity of writings assembles is not the author but the reader: 'the reader is the very space in which are inscribed, without any of them being lost, all the citations out of which a writing is made' (1986:54). In hypertext, the reader is fully engaged. To read a Xavier Herbert novel on a hypertext system that gives access to a diverse range of texts about twentieth-century Australia encourages wider-ranging questions. Instead of moving deeper into a single text, hypertext readers flit playfully between texts. Reading becomes a means of finding one's way from text to text. To immerse students in such a system is to encourage them to approach literature from multiple directions, and by 'that same associative model of understanding [which is] at the heart of Coleridge's notion of method' (Tuman, 1992a:66).

Hypertext also challenges the traditional view of literature as mimesis because an electronic author cannot hope to stabilise a replica of nature in so radically unstable a medium as hypertext. The relationship between author, text and nature is further complicated by the redefinition of the reader as an active participant in the making of meaning. The electronic medium 'denies the fixity of the text and . . . questions the authority of the author' (Bolter, 1991:153). In hypertext, two subjects—author and reader—combine in the text, and in a context made visible by the computer. Like Marxist critics, hypertext readers can learn how to read against the grain. It is difficult to read printed books subversively 'because the technology itself works against the reader's aggressive appropriation of the text' (ibid.). The computer, however, 'makes concrete the important act of reading (or misreading) as interpretation and challenges the reader to engage the author for control of the writing space' (ibid.:154). The computer fosters active reading habits that discourage readers from indulging in author-worship—or author denigration.

Being constantly in motion, the electronic text is never detached from the changing contexts in which readers place it. The desire to lose oneself in the world of the story—the goal of readers who read for entertainment—encourages passive reading. The reader of a hypertext, by contrast, is forced at every screen to reflect on the experience of reading. Reaching out to other texts, the electronic text invites readers to participate in its own construction. With hypertext, it is the reader who integrates the multiple and scattered parts into a whole. The process resembles what the anthropologist Claude Lévi-Strauss calls *bricolage*, which is the construction of something out of whatever materials are available. 'Every hypertext reader-author is inevitably a *bricoleur*', writes Landow (1992a:115): each reader-author constructs an individual text out of fragments. Such *bricolage* provides a new kind of unity that is entirely suited to hypertextuality.

Slatin (1990:875) suggests that 'hypertext systems tend to envision three different types of readers: the reader as browser, as user, or as co-author'. The browser wanders through an area, picking things up and putting them down 'as curiosity or momentary interest dictates' (ibid.). Browsers read for pleasure, and are therefore unlikely to go through all the material. While it is difficult to predict the pathways they will take, it is important to have a tracking mechanism that allows them to retrace their steps. The user is 'a reader with a clear—and often clearly limited—purpose' (ibid.:875). Users search for specific information, and leave the system when they have located it. Co-authorship is facilitated in hypertext when readers interact with the system to the extent of becoming 'actively involved in the creation of an evolving hyperdocument' (ibid.). Such interactions can result in either brief annotations or the creation of new links. They may represent a dynamic process

> in which the student moves among three different states: from a user the student becomes a browser (and may then become a user once again); ultimately, he or she becomes fully involved as co-author. Thus what looks like a hierarchy of readers collapses. (ibid.)

Hypertext invites readers to use, browse and co-author the text. Interactivity is highest when users begin to co-author the original text by creating new nodes and linking them to pre-existing nodes.

The capacity of hypertext to enable interactive reading and co-authorship represents 'a radical departure from traditional relationships between readers and texts' (Slatin, 1990:876). McLuhan's (1964) distinction between 'hot' and 'cool' media is relevant here: a cool medium invites active participation, whereas a hot medium assumes its audience is passive. Although McLuhan was thinking of the differences between print and television, it could be argued 'that hypertext combines the heat and visual excitement of film, video, and television with text's cool invitation to participate' (Slatin, 1990:876).

Hypertext offers more possibilities for the user-reader than earlier forms of software, and provides a context which invites playfulness:

> Armed with a bountiful hard disk upon which a 'classic' novel, poem, or play has been transcribed, the reader is free to become the producer of a new text by juxtaposing words, sentences, and paragraphs, creating new narratives and images, and scrambling the symbolic order of the 'original' author, revealing hidden dimensions of the work of art or treatise that had been statically enframed. (Aronowitz, 1992:133)

Such irreverent acts demonstrate that what we had thought of as permanent is really only contingent, and that high and popular cultures are 'ineluctably entwined' (ibid.) with technology. 'The assertions that the reader is the author and that the text is a system of signs whose meaning possesses no fixed centre have moved beyond the pretty formulations of a literary critic to become, at least tendentially, our new historical situation' (ibid.).

Although the 'rhetoric of hypertext . . . tends to be a rhetoric of liberation' (Bolter, 1992b:60), the reader's new-found freedom in hypertext can never be absolute. The goal of electronic writing may well be to set the reader free 'from all arbitrary fixity and stability of the print culture', but in fact, argues Bolter, 'hypertext simply entangles the reader in nets (or networks) of a different order', because readers are still bound by 'the constraints of the computer system and the constraints of the writing system the computer embodies' (ibid.). The computer system aims to be 'transparent' in its operations but of course it can never achieve that goal. Because readers must interact with a computer (by typing, moving the mouse or speaking into a microphone) they are obliged to know and obey the rules of interaction. The writing system may be any combination of words, graphics and video but 'as semiotics and postmodern theory in general have shown us, all such verbal and graphic writing must function in terms of codes and conventions. The reader can neither ignore nor circumvent these codes and conventions' (ibid.).

Disrupting the canon

Many people continue to believe in the enduring quality of the canon of great works of literature. Traditionalists such as Hirsch (1987) defend the canon as something with which every educated reader should be familiar. Contemporary literary theorists have critiqued the processes by which canons have been formed in western culture, and asked why certain works and authors have been included or excluded. Some object to the idea of a canon which is biased against minority and non-western writers, and dictates whom we must read. Scholes, for example, argues that

> the purpose of humanistic study is to learn what it has meant to be human in other times and places, what it means now, and to speculate about what it ought to mean and what it might mean in the human future. The best texts for this purpose

should be determined locally, by local conditions, limited and facilitated by local wisdom. Above all, they should not be imposed and regulated by a central power. (1986:116)

The critics of the canon question 'the whole complex of assumptions of . . . the traditionalists: the belief that there is a centre to our culture, that great works are unchanging in their message and importance, that great authors are authorities, and so on' (Bolter, 1991:151).

Traditional views of the canon and of reading, argues Bolter, 'were born of the technology of handwriting and matured in the technology of printing' (ibid.). If handwriting aspired to but could not achieve the stabilisation of the text, 'printing achieved a true cultural fixity, allowing texts to survive unchanged for centuries' (ibid.). Printing

> encouraged the making of canon in at least two ways. It ensured far more effectively than handwriting that a work could be 'timeless'—in the sense that it could survive for centuries without substantial change. And it provided dignity and distance for works of the canon: it made authors special by providing them with a writing space not available to other literate men and women. (ibid.)

A canon can be defined as 'a linear-hierarchical structure: it is a list of the most important authors and works, by which other, subordinate works can be measured' (Bolter, 1991:153). This makes it different from alternative structures such as a network. In a hypertext network, for example, all texts are joined by a dense texture of allusions and references. Some texts are more closely connected than others and therefore more rewarding; but in principle, a reader could begin anywhere and move by allusion and reference through the entire literary tradition. Electronic technology consequently 'threatens the idea of the literary canon' (Bolter, 1992a:36). Because all electronic texts are interrelated, none has well-defined borders; instead, each text reaches out to link up with past, present and future texts. It therefore becomes difficult to cordon off and to canonise a few great texts and authors. Some who oppose the notion of the canon envision those more fluid relationships between authors and readers that are posited in the electronic medium of hypertext. Here again, when contemporary literary theory speaks of empowering readers and breaking down the hierarchies of a traditional canon, it seems to be anticipating hypertext.

With the advent of hypertext, we 'lose a sense of the sanctity of the text' (Bolter, 1992a:35). The 'original' text no longer seems inviolate when readers are enabled to move through it, adding their own notes. 'Any text becomes a temporary structure in a changing web of relations with other, past and future, textual structures. In the culture that reads and writes electronically, the original text loses its privileged status' (ibid.:36). In contrast to printed books, electronic texts seem to offer the user energised control over even the literary classics. Lanham (1989:269), for example, imagines a student discarding a printed copy of *Paradise Lost* for a hypertext version, and then playfully adding music, graphics and graffiti, or altering the fonts. Accessing *Paradise Lost* on disk, the reader-writer can achieve a 'blasphemous rearrangement'. Lanham (ibid.) predicts that 'electronic media will change not only future literary texts but past ones as well'. Students should not hold literary texts in reverence, especially now that a new electronic technology makes it possible to interact with them. Conceived of as a mere text, even the most canonical of literary works is vulnerable in a system characterised by 'volatility, interactivity, easy scaling changes, a self-conscious typography, collage techniques of invention and arrangement, a new kind of self-consciousness about the "publication" and the "publicity" that lies at the end of expression' (Lanham, 1990:xiv–xv).

Traditional literary scholarship is heavily dependent on the idea that skilful editing can stabilise a printed text. Readers expect a printed text to represent its author's final words, and assume that the printed text of a canonical work carefully orders the constituent elements into the sequence in which they are to be read. Hypertext challenges in varying degrees such notions as stability and sequence. It threatens the privileged status of canonised works 'by unfixing them from their physical, unalterable status and placing them in the fluid medium of computer-based text' (Johnson-Eilola, 1994:210).

In hypertext, texts by students are no longer so easily subordinated to those by revered authors, and disparities between them are less visible. The text as a site of authority can also become a site of resistance: in hypertext, indeed, opposition to the canonised texts is more likely to succeed in conditions of hypertextuality than in the print culture, if only because hypertext makes it easier to expose the contradictions and power moves in such texts, and the multiply constructed positions from which they might be read. *Macbeth* translated into machine-readable form might be only one portion of an immense hypertext, in which students could add their own remarks to the mountain of commentary the play has received already. Hypertexts of this sort encourage readers and writers to produce those citations and quotations that in print texts, according to Derrida (1977:185–6), create 'an infinity of new contexts in a manner that is absolutely illimitable' because 'there are only contexts without centre or absolute anchoring'.

Even though hypertext may subvert the authority of the canon, its use may be limited, because many potential readers find it intimidating; certainly, its sheer complexity alone is enough to deter certain students. Furthermore, we cannot ignore the fact that electronic writing 'mediates its discourse through a technology whose expensiveness, complexity, and unreliability are still highly problematic' (Moulthrop, 1991c:265–6). But in spite of these limitations, electronic textuality accomplishes something that print can never manage: 'it unbinds writing' (ibid.:266). Once discourse has been organised into hypertext networks integrated into a decentred writing space, the possibility of open expression becomes more than a theoretical projection. Seeing that in fully developed hypertext any user is free to rearrange old links or create new ones, 'the old hierarchical model of textual authority must be rejected', according to Moulthrop (ibid.). Readers become not only writers but eventually publishers as well. Within such a system, the authority of the canon is disrupted irrevocably.

The new reading and writing

Despite important contributions by literary theorists in areas such as reader-response criticism and deconstruction, the roles of reader and writer have been construed as largely separate. Hypertext, by contrast, places reading and writing in a connected and overlapping terrain, thus providing 'a forum in which reading and writing can be reconceived in such a way that these traditionally separate acts begin to partially coalesce' (Johnson-Eilola, 1994:205). Even post-structuralist theories 'cannot fuse the reader and writer in the visible, surface-level manner experienced by hypertext reader-writers' (ibid.). In 'The Death of the Author', Barthes claims that 'in the multiplicity of writing, everything is to be *disentangled*, nothing *deciphered*; the structure can be followed, "run" (like the thread of a stocking) at every point and every level' (1993:117). And in 'From Work to Text', Barthes develops this idea by making each reader also partially a writer: 'The Text . . . asks the reader for an active collaboration' (1979:80). But whereas Barthes' transformation of reader into collaborative writer takes place only in the psychic world of the reader, 'hypertext makes the inter-textual, networked text visible and active for the reader-writer' (Johnson-Eilola, 1994:206). Furthermore, because the linking of texts or segments is so important in hypertext writing and reading, it may well encourage users to develop their own thought processes in this way, and 'to think in a pluralistic, nonlinear fashion' (ibid.).

Because its multi-linear approach destabilises the traditional roles of readers and writers, hypertext opens up new possibilities of enquiry. The creation of meaningful links among nodes of information results in dialogic interaction. The world of hypertext fosters the development of heteroglossia: 'everything means, is understood, as part of a greater whole—there is a constant interaction between meanings, all of which have the potential of conditioning others' (Bakhtin, 1981:426). Hypertext operators are neither solely readers nor solely writers but instead 'take the two roles simultaneously and visibly' (Johnson-Eilola, 1994:206). Hypertext not only invites readers to participate in making the text, 'but forces them to do so' (ibid.), thus requiring both readers and writers to become 'co-learners' (Joyce, 1995c:44).

Instead of viewing the writer as textual producer and the reader as textual consumer, we may consider the hypertext reader and writer as collaborators, each sharing in the process of mapping relationships among elements in the text. Hypertext invites its writer-readers to be both active and reflective by using the technology for reading and writing as well as for theoretical interpretation. Hypertext is 'a writing space of possibilities where readers interact with writers in a collaborative effort to make meaning' (Hager, 1993:103). Hypertext facilitates this meaning-making effort, which in its pure form allows readers and writers to engage in an ongoing dialogue, by allowing them both to write and to rewrite text. This enables hypertext to break with 'the old notion of singular texts and singular authorship' to foster instead 'plural texts and plural authorship' (ibid.).

The move from 'writer' or 'reader' to the 'writer-reader' of hypertext involves a negotiation and redistribution of traditional hierarchical power arrangements. With the loss of an authoritative and untouchable authorial identity comes a new sense of dialogic identity, created by the sense of being perpetually in dialogue with other texts and other writers. 'In hypertext, dialogues between students, teachers, and canon texts and authors may be even more egalitarian, because . . . in this medium, there is no way to resist multiplicity' (Johnson-Eilola, 1994:213). The structure that facilitates such dialogues is the network.

The changes effected by hypertext invite us to redefine reading and writing. In hypertext, the common distinction between 'writer' and 'reader' begins to collapse in a way that has long been theorised for printed text but never before realised in such a visible form. How people read and write in the print medium differs in important ways from how hypertext encourages them to read and write. We are familiar with the structures of traditional text, but still lack theories of how to deal with hypertexts, especially those that cultivate complex linkages within and between a complex set of texts.

Attempts to articulate a theory appropriate to the changes wrought by hypertext on reading and writing practices are now beginning to emerge. Ulmer (1989), for example, has developed the concept of 'applied grammatology'. Unsure whether Derrida has already heralded or merely glimpsed the new age of writing, Ulmer takes Derrida's ideas further. For Ulmer, we are witnessing a communications revolution that has deep cultural roots: it mixes images and words in new ways, and (as in an oral society) substitutes narrative for conceptual argumentation. Academic discourse, which perpetuates the myth that it is possible to separate fiction from truth, has abandoned the way people think in the world outside the academies. In order 'to provide the mode of writing appropriate to the present age of electronic communications', Ulmer (1985:303) has developed his notion of 'applied grammatology', which, he believes, provides for writing in the age of information a theoretical frame that is free of an ultimate commitment to book culture.

Electronic culture, Ulmer argues, will be as different from the culture of the book as book culture is from oral culture. For one thing, it enables all three dimensions of discourse—oral, print and electronic—to coexist interactively. Ulmer (1992:140) finds Derrida's figure of *mise en abyme* very interesting in the context of hypertext. 'The *mise en abyme* is a reflexive structuration, by means of which a text shows what it is telling, does what it says, displays its own making, reflects its own action' (ibid.). Ulmer's 'Grammatology (in the Stacks) of Hypermedia' is an experimental attempt 'to use the *mise en abyme* figure to organise an "analysis" of the current thinking about hypermedia' (1992:141). His strategy is to imitate in alphabetic style the experience of hypertext: navigating through a database, and producing a trail of linked items of information. Ulmer adopts the stack format of Hypercard, confining himself to citations from a diverse bibliography of materials relevant to hypertext.

Ulmer's resultant hypertext essay consists of twenty-nine cards which simulate one trail blazed through a domain of information about hypertext. To Ulmer, hypertext may be 'the technological realisation of Walter Benjamin's dream—a book composed entirely of quotations. As in Benjamin's Arcades project, the simulation is a collage, relying upon the remotivation of preexisting fragments in a new context for the production of its own significance' (1992:142). It represents, perhaps, a new genre of academic discourse, what Ulmer calls 'mystory'. A 'mystory' combines elements of personal autobiography, oral history, and expert conceptual analysis. '"Mystory" puns on "mystery"', writes Ulmer: 'a good mystory should surprise us by its juxtapositions and changes in declaratory and stylistic register' (1985:63). A collage, it claims originality only from the particular pattern of juxtaposition it offers. The implications of Ulmer's notion of collage—a mode of academic discourse particularly suited to the humanities and social sciences, and one which embodies most elegantly what it is that hypertext does—are explored in Chapter 6.

Reconceiving Narrative

Hyperfiction and literary theory

When writers use hypertext to produce a fictional narrative, the result is a new literary form: interactive hyperfiction. A significant distinction between traditional print narratives and hyperfiction lies in how we approach them. Readers of print narratives usually begin on the first page and, even though they may move backwards or forwards, generally proceed through the text to the end. Their gradual progression follows a carefully scripted route which ensures that they get from the beginning to the end in the way the author wants them to. By contrast, most hyperfictions have no single beginning or end. A further distinction is based on the tangibility of the text. Whereas the length of a work of fiction can be gauged just by holding it, readers of a hyperfiction do not know what the hypertext contains till they load it into their computer, and even then they may never experience its full magnitude. However, 'in the physical intangibility of interactive narratives lies one of the keys to their flexibility. Because the narrative is not fixed and locked into place in typeset lines, readers can interact with the story in what they choose to read' (Douglas, 1992:4). The possibilities for readers to create their own stories are considerably greater in hyperfiction than when reading a print narrative or Choose-Your-Own-Adventure, both of which have highly visible beginnings and endings, as well as other structural limitations.

In discussions of the impact of hypertext on literary form, two different approaches are common. In the first, hypertext is 'used as a lens' to reveal what was 'previously unnoticed or unnoticeable'. Theorists identify 'quasi hypertextuality' (Landow, 1992a:102) in print texts such as Lawrence Sterne's *Tristram Shandy* (1759–67) and James Joyce's *Ulysses* (1922) which 'reveal new principles of organisation or new ways of being read to readers who have experienced hypertext' (ibid.) and can appreciate the degree to which both authors produce texts that resist linearity. Other theorists, however, deduce the qualities of hypertext narrative 'from the defining characteristics of hypertext—its non- or multilinearity, its multivocality, and its inevitable blending of media and modes, particularly its tendency to marry the visual and the verbal' (ibid.: 103). Both approaches are discussed in the following sections.

Literary and electronic precursors

Although books are a poor medium for participatory discourse, since the beginnings of modern fiction authors have attempted 'to jar or cajole readers out of passivity' (Kaplan and Moulthrop, 1991:11). Literary precursors of hyperfiction include not only *Tristram Shandy* and *Ulysses* but also more recent fiction such as Julio Cortazar's *Hopscotch* (1966) and Borges's 'The Garden of Forking Paths'. As unique solutions to the linearity of conventional fiction, such 'narratives . . . explore, exploit, and ultimately chafe at the confines of printed space' (Douglas, 1992:13). Sterne, Joyce, Cortazar and Borges are self-consciously absorbed in both the act of writing itself and the difficult relationships between narrator, text and reader. All work strenuously against the medium in which their books are produced. Any reader familiar with hypertext will look at such texts anew, and observe that in their resistance to linear narrative they have much in common with hyperfiction. By attacking the convention that a novel is a coherent narrative of events, such texts simultaneously invite and confirm reader-interaction.

In *Tristram Shandy,* Sterne dislocates and distorts the order of events in an otherwise simple story to create a complex plot. Tristram, as narrator, is always interrupting the story to remind us that he is in fact writing his life history, and to draw our attention to various digressions and omissions. Tristram explains that 'when a man is telling a story in the strange way I do mine, he is obliged continually to be going backwards and forwards to keep all tight together in the reader's fancy' (Sterne, 1987:370). Sterne also subverts the conventions of printed or written texts when he not only suggests to his readers that a page is missing but also leaves a blank space on which they themselves are invited to write a few words in response to his text. Tristram thus involves his readers in the very making of the book. Omissions in the text are indicated by asterisks: the substitution of one asterisk for each letter, for instance, creates a code for the reader to decipher. Such games challenge readers to write the printed text.

But whereas Sterne can only pretend to offer his readers the opportunity to take part in the construction of his book, hypertext can demand that the reader participate. In a hyperfiction, no text appears on the screen until a reader summons it with a keystroke or the click of a mouse. Furthermore, argues Bolter, the electronic environment gives a stronger sense than does the printed page of the author 'being there'. 'The author is present in the electronic network of episodes that he or she creates and through which the reader moves along associative paths' (Bolter, 1991:134).

Within the constraints of the print medium, *Tristram Shandy* instantiates the possibility for readers to participate actively in the events of the hero's life. *Ulysses,* on the other hand, embodies the tension between our linear experience of reading and the novel's allusive structure. The book is about a day in Dublin and the meeting of Stephen Daedalus and Leopold Bloom; it is also a pattern of allusions to previous literary and cultural texts. The narrative carries its readers forward from one incident to the next on that day in June 1904, while the allusions entice them in other directions, and notably to Homer's *Odyssey. Ulysses* is in this respect a palimpsest, in which Joyce's text is written 'over' Homer's: we see and are meant to see both, but we cannot distinguish older from newer layers of textuality because the final printed edition is seamless. What readers find is 'a self-referential text': they move 'back and forth through the book in order to appreciate the complex relationships of its parts' (ibid.:135). A hypertext version of Joyce's text would allow readers, as with a palimpsest, to uncover the underlayers of allusion. In the electronic medium, Joyce's text would be reconstituted as a massive and interactive network that readers could 'traverse in a variety of ways' (ibid.:137) in pursuit of many possible readings.

In contrast to the novels of Sterne and Joyce, Cortazar invites us to read *Hopscotch* 'in a successive, literally programmed iteration' (Joyce, 1991a:83). This strategy—used in Choose-Your-Own-Adventure books—allows the reader to produce a number of divergent story versions, each mutually exclusive. It should theoretically be possible to produce an extremely variable text in this way. But even when page sequencing is manipulated with great sophistication, as in Cortazar's *Hopscotch*, the reader is free to choose from only a small range of variations on a discernibly central theme. Cortazar, after insisting that his novel consists of 'many books', outlines in fact only two recommended plans for reading it.

Even though *Tristram Shandy*, *Ulysses* and *Hopscotch* create multiplicities as intricate as any of those envisioned for interactive fictions, they 'were bound at very least by the static nature of their presentational systems' (ibid.:82). Hypertext, on the other hand, promises fiction writers 'a means to resurrect and entertain multiplicities that print-bound creation models have taught them to suppress or finesse' (ibid.:81–2).

Hyperfiction develops from a specifically twentieth-century tradition of experimental literature. Dadaism, for example, aimed at destroying the structures of established art and literature, and 'in that breakdown the Dadaists worked in the same spirit as writers now work in the electronic medium' (Bolter, 1991:131). Dadaists such as Hans Arp and Hugo Ball often attacked the conventions of the realistic novel that tells its story with a clear and cogent rhythm of events, and in doing so found themselves straining at the limitations of the printed page. Because the linear-hierarchical presentation of the printed book was so well suited to the conventions of plot and character in the realistic novel, 'to attack the form of the novel was also to attack the technology of print' (ibid.). The *nouveau roman* in France, like 'programmed' and 'aleatory' novels elsewhere, aimed at subverting the conventions of printed literature.

The idea of the cut-up or collage method also had its origins in Dadaism. In the 1920s Tristan Tzara developed a technique for writing poetry by chance rather than by choice: 'he would jumble up some words written on slips of paper, and take them out of a hat, one by one to see what poem resulted' (Woolley, 1992:155). William Burroughs, whom some regard as a proto-cyberpunk novelist, explains his compositional technique in similar terms:

> The method is simple. Here is one way to do it. Take a page. Like this page. Now cut down the middle and across the middle. You have four sections: 1 2 3 4 . . . one, two, three, four. Now rearrange the sections placing section four with section one and section two with section three. And you have a new page. Sometimes it says much the same thing. Sometimes something quite different. (Calder, 1982:19)

Burroughs developed this technique in order to destroy the control of narrative by sequentiality: why should *z* follow *y*, day follow night, two follow one? These apparently 'natural' sequences, according to Burroughs, are just one set of progressions among many.

Other twentieth-century novels, plays and films also critique narrative conventions. In *The French Lieutenant's Woman* (1969), John Fowles gives the book two endings (or three, if you count the one that occurs three-quarters of the way through, which supplies a conventional Victorian outcome). Fowles here highlights the spurious meaningfulness of the fictional world he has created and the indeterminacy of the real one. In Karel Reisz's and Harold Pinter's 1981 film of the novel, the two endings are translated into a film-within-a-film. In his related series of plays, *Intimate Exchanges* (1985), Alan Ayckbourn provides a chart of the consequences stemming from a single opening scene. 'Sixteen divergent conclusions sprout from eight main events derived in turn from four sessions of enlightenment, two introductory encounters and one crucial moment of decision' (Strick, 1994:47). Alain Resnais' film version of Ayckbourn's play, *Smoking/No Smoking* (1993), dispenses with this permutational guide and leaves it to viewers to make sense of the unpredictable twists in the lives of the characters. It comprises two contiguous films, in which the action proceeds 'in a staccato of climactic fade-outs, each followed by a renewed build-up' (ibid.); it concludes with twelve final graveyard scenes, 'settings for a perpetual litany, a barely-concealed scaffolding, of christenings, marriages, funerals, thanksgivings and remembrances' (ibid.:48).

Because interactive fiction already existed in print and film, the technological challenge for creators of electronic interactive fiction was 'to find a way of turning imaginary worlds lodged in the writer's head into virtual worlds lodged in the computer's memory' (Woolley, 1992:155). The precedent was Adventure, developed in the 1960s at Stanford University's Artificial Intelligence Laboratory (SAIL). The program was conceived of as an experimental game. A computerised version of role-playing games like Dungeons and Dragons, Adventure comprises a series of descriptions of fictional locations inspired by J. R. R. Tolkien's fantasy *The Lord of the Rings* (1954), and set in the surrounding Californian mountains. It maps an imaginary environment into electronic memory and allows its player-readers to explore that space by issuing simple commands (e.g., go north, get flashlight, kill troll with axe). In giving these commands, the reader attempts to negotiate a series of spatial and narrative obstacles to reach some hidden goal. Adventure became a diversion of programmers and computer scientists, 'who built ever more intricate and challenging versions of the game' (Kaplan and Moulthrop, 1991:12).

Adventure and its descendants continued to evolve through the late 1970s, when interactive text games migrated from academic and corporate mainframes to home computers. There the form was married with popular fiction and role-playing games to produce a second generation of text adventures that retained the problem-solving design of the original Adventure. These games were not networks of possibilities to be explored but arrangements of obstacles to be overcome in the progress to a determined goal. Later in the 1980s there emerged a third generation of interactive fiction in which the influence of game scenarios has been less noticeable. The multiple fictions of this third generation are narrative networks capable of differing significantly on every reading. They 'do not vector the reader toward a single closure or solution but enable a multitude of outcomes' (Kaplan and Moulthrop, 1991:13).

Hyperfiction

In interactive hyperfiction, the reader determines the story's outcome by controlling its branching of events. Existing hyperfictions resemble two of the divergent modes explored in avant-garde or experimental fiction: 'narratives of multiplicity' and 'mosaic narratives' (Douglas, 1992:16). Pre-hyperfictional narratives of multiplicity such as Borges's 'The Garden of Forking Paths' and Fowles's *The French Lieutenant's Woman* use the linear and sequential nature of print to explore simultaneity and multiplicity in plotted events. Although typographic conventions oblige each of these plots to unfold separately and in order, narrative possibilities and conclusions proliferate. But whenever they try to withstand the physicality of print by 'increasing the number of stories, narrative strands and potential points of closure . . . the medium inevitably resists, making the reading experience and the significance of the narrative itself more a meditation on the confines of print space than anything else' (ibid.:19).

Mosaic print narratives, such as Lawrence Durrell's *The Alexandria Quartet* (1957–60), Cortazar's *Hopscotch* and Barthes' *The Pleasure of the Text* (1975) 'consist of narrative fragments, conflicting perspectives, interruptions, and ellipses which impel their readers to painstakingly piece together a sense of the narrative, with its full meaning apparent only when viewed as an assembled mosaic, a structure embracing all its fragments' (ibid.:16). *The Alexandria Quartet*, which in four separate volumes reveals different aspects of the same events, perhaps comes closest in the print medium to presenting its readers with the discrete and self-contained narrative perspectives that they might encounter in such hyperfictions as Afternoon and Forking Paths.

Joyce's hyperfiction Afternoon (1991b) is an intricate web of narratives, places, paths and 'yields', that is, words and phrases whose evocative resonances readers can pursue by using a mouse to highlight them on the computer screen. Afternoon is a fiction that changes every time it is read. It invites the reader to circulate digressively among a matrix of characters and events that are never quite what they seemed on first presentation. 'I want to say I may have seen my son die this morning', an anonymous speaker confides, disclosing a rich field of narrative possibility. However, none of the stories produced by interacting with *Afternoon* will validate or disprove either the desire or the perception of the speaker.

Afternoon is a text scattered with verbal associations. If you select the word 'son' in the first sentence of the story, for example, the text on the screen shifts to a description of the scene in which the narrator, who appears to be male, finds his son's school essay 'The Sun King' on Louis XIV. The word 'die' in the initial sentence serves as the cue for a different narrative departure. But there is also a third possibility—a default condition—activated when you select any words other than those singled out as special operators. These default transitions, however, do not simply reinstate the fixed page-order of a bound volume. Governed by conditions that refer to a record of your encounter with the text, they are subject to change. *Afternoon* is structured in such a way that its elements are assembled in a different order every time you call up a new screen.

Hyperfiction fosters both passive and active reading, or what Bolter calls 'looking at and looking through' the text (1992a:40). When reading an episode, you may succeed in looking through the text to an imagined world. Formal structures are both visible and operative in hyperfiction because they are embodied in the links between episodes. At each link the text offers a series of possibilities that you can activate, moving backwards and forwards between the verbal text and the structure as you read. In *Afternoon* you may get lost in Peter's engaging story of his search for his son. But the need to make choices never lets you forget that you are participating in the making of a fiction.

There is no plot as such in *Afternoon*. Because it is not built on causal sequences, it does not present parallel story-lines. Events are ambiguous, and the story focuses on how the characters might react to such ambiguities. There appears to be a mystery: the narrator's son may or may not have been in an automobile accident. Readers are compelled to follow the father as he tries to establish the fate of his son: in this respect, the father's quest becomes the reader's. The particular episodes you call up will determine the answer you receive. 'In that sense, "Afternoon" becomes an allegory of the act of reading', Bolter observes. 'The reader's own participation in the story becomes the story' (ibid.:29).

Afternoon differs from printed fiction by not offering any 'single story of which each reading is a version, because each reading determines the story as it goes'; as a result, 'there is no story at all; there are only readings' (Bolter, 1991:124). We could also say that the story of Afternoon is the sum of all its readings, in so far as the story is a structure that can embrace contradictory outcomes. 'Each reading is a different turning within a universe of paths set up by the author' (ibid.:124–5). Douglas calls Afternoon 'a sort of computer-driven version of Borges's *Book of Sand*' (1991:117)— an infinite text which never offers the same page to any reader more than once.

Michael Joyce's Afternoon demonstrates that it is possible to create a text which does not force its readers down one particular route. A corollary is that readers risk becoming lost, partly because the textual landscape is unfamiliar and partly because the narrative is the means by which readers orient themselves. Joyce recognises such difficulties and seeks to overcome them by placing limits on narrative freedom, although in Afternoon he does not provide his readers with a map. In WOE, however, Joyce (1991c) includes a map of the text's overall structure and of places still awaiting discovery. It records previous paths and suggests which directions might prove fruitful for exploration.

Afternoon and similar hyperfictions are quite different from the kinds of experiments that Burroughs made with printed texts. For example, Afternoon is not an aleatory, random fiction because its author exercises control over the choices his reader can make. Afternoon can be (and sometimes is) a linear story, because occasionally only one path leads from an episode. At other times, it gives its reader dozens of choices, although they are 'far from random, and . . . do not impress the reader as haphazard' (Bolter, 1992a:29). As a text that changes before our eyes, Afternoon challenges our assumptions about the nature of literature: it represents 'a new kind of writing' (ibid.). But because it also comes out of a literary tradition, we recognise it as a coherent act of imagination, as a story with characters who interact and conflict. Yet because it is an electronic text, Afternoon 'is not meant to be read, but always reread' (ibid.:30). Perhaps we may never break free of the linear mode of reading. Nevertheless, links between elements in a network release us from that single ordering of material which the first reading of a printed text requires.

Whereas Afternoon is Joyce's own creation, Moulthrop's (1986) Forking Paths is an adaptation of Borges's 'The Garden of Forking Paths' (1970c). At the centre of Borges's story is a description of a Chinese novel that seeks both to explain and to defy time. It tells how the author of the novel, Ts'ui Pen, was thought to have retired from public life with two objectives: to write a book and to build a labyrinthine garden. The sinologist Stephen Albert discovers, however, that these two goals were really one: that the book *was* the garden. The manuscript that Ts'ui Pen left behind was not, as it seemed, 'an indeterminate heap of contradictory drafts' (ibid.:50), but rather a tree of all possible events. Albert explains:

> In all fictional works, each time a man is confronted with several alternatives, he chooses one and eliminates the others; in the fiction of Ts'ui Pen, he chooses—simultaneously—all of them. *He creates*, in this way, diverse futures, diverse times which themselves also proliferate and fork. Here, then, is the explanation of the novel's contradictions. (ibid.:51)

Albert tells us that the Garden of Forking Paths is a game whose subject is time. Ts'ui Pen believed 'in an infinite series of time, in a growing, dizzying net of divergent, convergent and parallel times. This network of times which approached one another, forked, broke off, or were unaware of one another for centuries, embraces *all* possibilities of time' (ibid.:53). The metaphor of the garden suggests 'a luxuriant growth of textual possibilities' (Bolter, 1991:139). Both that suggestion and the story itself are suddenly closed off when Albert himself is murdered by the narrator of the story. The abrupt ending contrasts markedly with the novel Albert describes, which refuses to select one of its bifurcating paths in order to reach a single end. Borges's story suggests that the end of a text is always arbitrary.

The options in Moulthrop's hypertext, Forking Paths, are not arranged hierarchically. Narratives branch out from both the characters' names and possible scenarios suggested in the short story; all are potentially available for discovery, development, and conclusion in this interactive hyperfiction. Readers who are not satisfied with a single conclusion can continue reading after discovering that a stranger, Yu Tsun, has murdered Stephen Albert. Returning to the text, they may find that 'the hitherto peaceable Stephen Albert has unexpectedly garrotted Yu Tsun' (Douglas, 1989:97). Moulthrop offers twelve permutations on the ending, as well as retellings of strands from alternative points of view, unexpected reversals in character traits and motives, and occasionally a playfully metatextual commentary on the nature of interactive text itself. Moulthrop capitalises on one of the most obvious potentials for interactive narrative: its ability to provide readers with the option of choosing which narrative scenarios to explore and resolve. Forking Paths thus 'promises multiple, although not infinite, differences in a number of readings of the narrative' (ibid.).

Whereas 'mosaic' print narratives may give their readers greater autonomy in allowing them to prefer certain developments or conclusions to others, hyperfictions like Forking Paths present discrete pieces of information that require a participatory act on the part of the reader. Bits of experience have been textualised and segmented by the author, but readers must decide for themselves how it is to be reassembled. In this respect, Forking Paths is like Barthes' *The Pleasure of the Text* and Cortazar's *Hopscotch*. The order in which these mosaic texts are configured relies less upon established narrative conventions than upon the connections realised by each reader. 'From the reader's point of view, an interactive fiction very closely resembles Iser's (1980) virtual work. Its order surfaces (so it seems) only in the act of reading' (Kaplan and Moulthrop, 1991:12).

Unlike interactive print fiction, hyperfiction abandons such conventions as chapters and the illusion of a seamless continuity between paragraphs. The virtual text exists only in electronic space or in our memories. 'There is . . . no single story and, contrary to our expectations based on reading print narratives, readings do not provide varying versions of this story or collection of stories' (Douglas, 1992:14). Because it generates or determines the story as it proceeds, 'each reading is a different turning within a universe of paths set up by the author' (Bolter, 1991:124–5). As the text re-forms with successive readings, no two readings are alike. By hyperfiction presenting a chameleon text-like surface, hyperfiction is textually subversive. Its structure is 'effectively open to a virtually unlimited range of possible readings, each of which causes the work to acquire new vitality in terms of one particular taste, perspective, or personal performance' (Eco, 1979:63).

But hyperfiction also arouses unease if not antagonism in some users. In her initial encounter with the technology, Douglas was enthusiastic about interactive fiction in hypertext form. A year later, she had qualified her excitement:

> Although we may already have examples of fiction which suggest an impatience with the restrictions and confines of the print narrative and a readiness to circumvent or even transcend them, we also have an electronic environment alien and inimical to our habitual reading patterns. (1989:94)

Interactivity

The notion of interactivity in fiction has been with us for much of this century: readers are widely seen as 'breathing life into the texts they read' (Douglas, 1992:9). To read a print narrative, however, is far from a *literally* interactive exchange between oneself and the author, if by 'interactivity' we mean something like what happens in conversation when two or more people engage verbally with one another. By comparison, any reader engrossed in Christina Stead's *The Man Who Loved Children* (1940) is positioned like someone listening to a monologue. You can interrupt only by closing the book or allowing your attention to wander. Except in the case of experimental writing, there is only one path through a print narrative. If you try to focus exclusively on the references to a particular aspect of *The Man Who Loved Children*—such as Henny's recollections of her childhood—

attempts at interaction are likely to collapse into incomprehension. Even if you skip back and forth through the text, the grim ending of Stead's book is invariably the same. Moreover, the text would always be confined to the printed words between two covers, even if its 'repertoire of interpretative strategies . . . were to embrace the entire, existing literary canon' (Douglas, 1992:11).

By contrast, readers who 'open' Moulthrop's hyperfiction Victory Garden (1991d) are obliged to interact with it before they can enter the narrative. The first three sites they encounter are made up of lists of other places and paths, from each of which they must make a choice in order to begin navigating through the narratives in the hypertext. Although these lists resemble a table of contents—'Places to be', 'Places to explore' and 'Paths to deplore'—they 'do not represent a hierarchical map of the narrative' in the sense of 'providing readers with a preview of the topics they will explore . . . and the order in which they will experience them' (Douglas, 1992:3). The 'first' place or path in a list has no priority over the others: 'readers will not necessarily encounter it first and need not encounter it at all in the course of their reading' (ibid.). Each of the listed words or phrases acts instead merely as a contact point for entering the narrative. 'By choosing an intriguing word or particularly interesting phrase, readers find themselves launched on one of the many paths through the text' (ibid.). In a printed novel there is no table of contents. If there is it is generally irrelevant to our reading experience, which 'begins with the first words of the narrative and is completed by the last words on the last page' (ibid.).

A hyperfiction can also provide an overview of its narrative structures. Presented as a flat and schematic image, it nevertheless seems more like a holographic representation which readers can move around or even through. Indeed, in hyperfiction, 'the burden of interactivity and the continual necessity to choose directions for movement never allows us to forget that we are reading by navigating through a "space" which contains length, depth and height' (ibid.:14), and through which we move effortlessly and instantly from one place to another. By contrast, the spatial relations or spatial form encountered in conventional print narratives are like those in cinema, where 'we see three-dimensions represented and projected on a flat, one-dimensional plane' (ibid.:13).

Douglas (1991:112) compares the experience of navigating Joyce's WOE (1991c) with that of a printed book. 'Ordinarily, reading is virtually undetectable, apart from the delicate twitchings of the eye and the odd subvocalisation here and there'; when you open a reading log in WOE you can see your 'movements crystallised'. WOE offers its readers an overview of the text's structure and the places to be discovered. This cognitive map records previous wanderings and suggests which directions to explore. For Douglas (ibid.:124), this represents 'a move toward a degree of interactivity inconceivable in print narratives', which require readers not only to construct a chronology and map of the actions unfolding in the novel but also to integrate their understanding of the plot with the order in which it is revealed in the narrative. What results is 'a sort of reading in two dimensions' (ibid.). WOE takes reading into a third dimension; for by using the cognitive map, readers 'can actually see vertical and horizontal relationships between elements in the text' (ibid.:124).

Of course, neither conventional nor electronic texts can be wholly interactive, 'since true interaction implies that the user responds to the system at least as often as the system responds to the user, and, more importantly, that initiatives taken by either user or system alter the behaviour of the other' (Joyce, 1991a:80). Although interactive fiction provides a much more engaging simulacrum of dialogue than does print, it is still merely a simulation. However 'personalised' the reader's path may be, it is always contained within a predefined matrix of possibilities. Afternoon certainly constrains its readers less than an adventure game does, but it offers them nevertheless only a limited choice of 'yields'.

Exploring narrative form

We love a good story, told by a skilled narrator, and dictated by the authoritative voice of an accomplished author. Indeed, the lure of narrative is both powerful and enduring. Hypertext invites us 'to find an analogue in the electronic medium for narrative line and authorial control in the traditional medium of print' (Bolter, 1993:9). The most effective techniques for achieving a strong story-line in the print medium are linearity, plot, characterisation, textual coherence, resolution and closure. Experiments in hyperfiction, however, diminish these qualities in varying degrees by exploiting the electronic medium's capacity to create open-ended fictions with multiple narrative strands. Any discussion of the changes to narrative form brought about by hyperfiction necessarily involves a consideration of the ways in which writers using hypertext technology have played with these integral elements, and found alternative strategies and techniques for engaging readers' attention.

In one sense, each reading of a hyperfiction is a linear experience: confronted with one frame after another, you are still aware of a narrative, however confused it may be. At the same time, a hyperfiction seems to contain more than one voice and to change direction abruptly. Each hyperfiction handles in its own way the conflict between the linearity of the reading experience and the multiplicity of hyperfiction. In Joyce's Afternoon, for instance, some readings represent alternative voices or perspectives on the narrative, with each discrete change kept separate by electronic space. The web of intersecting narrative strands in Moulthrop's Victory Garden offers a mixture of voices and genres: first- and third-person narrative fiction, excerpts from other books, fiction and non-fiction, and quotations from televised broadcasts. Joyce's WOE is a narrative experiment in which some readings are metafictional commentaries on the narrative and its origins in Joyce's experience.

In hyperfictions such as these, a lack of linearity does not destroy the narrative. In fact, since readers always fabricate their own structures, sequences and meanings—and particularly so in hypertext conditions—they have surprisingly little trouble constructing a story as they make their way through the web. Reading hyperfiction, however, can be a very different experience from reading a printed novel or a short story—so much so as to put us in mind of that new 'orality' which both Marshall McLuhan (1964) and Walter J. Ong (1982) predicted would develop with the demise of print culture. What hyperfiction forces us to recognise is that an active author-reader fabricates not only meanings but also a text from the kit supplied by the author.

At first, readers of hyperfiction may be reluctant 'to think of a text as at once immutable and ever-changing' (Amato, 1991:112–13). Yet interactive hyperfiction encourages 'a more disjunctive, less linear, more casual (hence less causal), [and] ostensibly more open-ended textual experience' (ibid.:113), provided that readers are willing to modify their expectations of 'text'. Both readers and writers of hyperfiction need to extend the range of traditional narrative in order to forge a writing space appropriate to the electronic medium.

Reconceptualising plot and story

In his *Poetics*, Aristotle offers a definition of plot in which sequence plays a central role. A narrative must have a beginning, a middle and an end: 'well plotted fables must not begin or end casually, but must follow the pattern here described' (Aristotle, 1959:27). Hyperfiction interrogates not only Aristotelian notions of 'beginning' and 'end', but also his assumptions about the 'sequence' of parts and the 'unity' of the finished work. If we accept Landow's argument, as outlined in Chapter 3, that hypertext is an effective means of testing literary theory, then hyperfiction calls into question some of the most basic points about plot and story in the Aristotelian tradition.

Hyperfiction apparently dispenses with linear organisation: linearity becomes 'a quality of the individual reader's experience within a single lexia and his or her experience of following a particular path, even if that path curves back upon itself or heads in strange directions' (Landow, 1992a:104). Although the experience of linearity does not disappear altogether with hyperfiction, narrative chunks do not follow one another in a page-turning, forward direction. Hyperfiction space is multi-dimensional and theoretically infinite: its set of possible network links are fixed, variable or random. Readers can contribute by choosing their own route through the labyrinth; the more active may introduce new elements, open new paths, and interact with the characters or even with the author(s). In this electronic space, writers need a new concept of structure:

> In place of a closed and unitary structure, they must learn to conceive of their text as a structure of possible structures. The writer must practise a kind of second-order writing, creating coherent lines for the reader to discover without closing off the possibilities prematurely or arbitrarily. (Bolter, 1991:144)

Parataxis, 'which is produced by repetition rather than sequence' (Landow, 1992a:106), is one way of organising narrative in hyperfiction. In works that depend upon a logical or temporal organisation, the sequence as a whole can be rendered incomprehensible (or its effect can be radically altered) if a particular element is dislocated or omitted. Paratactic structures, however, are not generated causally, and therefore one element need not follow from another: consequently, 'thematic units can be added, omitted, or exchanged without destroying the coherence or effect of the . . . thematic structure' (Smith, 1968:99). Paratactic structures can be formed by either variations on a theme or by a list. The fact that a paratactic structure cannot determine its own conclusion is not so much a problem as a strength. The absence of closure or resolution is exactly what attracts many theorists to hypertext. On the other hand, if, as some narratologists claim, 'morality ultimately depends upon the unity and coherence of a fixed linear text' (Landow, 1992a:106), then hyperfiction may be criticised as a morally bankrupt and corrupting form of postmodern fiction.

The existence of earlier experiments in avoiding the linearity of the printed text suggests that many authors have felt that it falsified their experience of things. Tennyson, for example, 'created his poetry of fragments in an attempt to write with greater honesty and with greater truth about his own experience' (Landow, 1992a:107). Non-linear narrative may represent our perceptions more accurately. Many twentieth-century works of fiction explore the tension between linearity and a more spatial sensation of time. Patrick White's *Riders in the Chariot* (1961) and Janette Turner Hospital's *The Last Magician* (1992), for example, question the status of sequence in narrative and so too does David Malouf's *Fly Away Peter* (1982). The protagonists of all three novels are suspicious of chronology and sequence: what they experience is simultaneity, as revealed from the perspective of a kaleidoscope. According to Paul Ricoeur, 'the major tendency of modern theory of narrative . . . is to "dechronologise" narrative': White, Hospital and Malouf each conducts an effective 'struggle against the linear representation of time' (1984:30). The difference between their novels and hyperfictions is that hypertext confers greater freedom and power on the reader. Malouf decides at what point his protagonist's narrative is to branch out; in Moulthrop's Victory Garden and Joyce's Afternoon, the reader makes that kind of decision.

Reconceptualising beginnings and endings

The problems posed by hyperfiction for traditional understandings of narrative 'appear with particular clarity in the matter of beginning and ending stories' (Landow, 1992a:109). If 'beginnings imply endings, and endings require some sort of formal and thematic closure' (ibid.:110), what is the significance of such terms in the context of hyperfiction? Ricoeur explains that 'to follow a story is to move forward in the midst of contingencies and peripeteia under the guidance of an expectation that finds its fulfilment in the "conclusion" of the story'. This is what 'gives the story an "end point", which, in turn, furnishes the point of view from which the story can be perceived as forming a whole'. To understand a story thus requires grasping 'how and why the successive episodes led to this conclusion, which, far from being foreseeable, must finally be acceptable, as congruent with the episodes brought together by the story' (1984:66–7). Literary convention decrees that endings must either satisfy or in some way respond to expectations raised during the reading of the narrative.

In their brief history, hyperfictions seem to have taken 'an essentially cautious approach' (Landow, 1992a:109) to the problem of beginnings by offering the reader a block of text—labelled with something like 'start here'—that combines the functions of title page, introduction and opening paragraph. There are various reasons for this. One is convenience: the disk has to be self-contained so that it can be used on stand-alone machines. Another is the reluctance of some writers to disorient readers at the point of their first contact with the narrative. A further reason is that some believe hyperfiction should change our experiences of the middle but not the beginning of narrative fiction. The rival view is that because hyperfiction uniquely enables us to begin with any one of its parts, we should take advantage of this fact. In order to achieve this end, each chunk of text must be sufficiently independent to generate meanings that can be followed in other chunks of hyperfiction.

Although they use familiar narrative strategies to make beginnings easier, hyperfictions challenge readers by avoiding the corresponding devices for achieving closure. It is up to readers to decide how, when and why the narrative finishes. In Afternoon, Joyce makes closure the responsibility of the reader. In a section entitled 'work in progress' we are advised that 'Closure is, as in any fiction, a suspect quality, although here it is made manifest. When the story no longer progresses, or when it cycles, or when you tire of the paths, the experience of reading it ends'. Joyce, however, continues:

> Even so, there are likely to be more opportunities than you think there are at first. A word which doesn't yield the first time you read a section may take you elsewhere if you choose it when you encounter the section again; and what sometimes seems a loop, like memory, heads off in another direction.

Afternoon 'has so many points of departure within each lexia as well as continually changing points of linkage [that] one sees what Joyce means', observes Landow (1992a:113). Joyce intertwines death and narrative so closely that if you decide to stop reading Afternoon you kill the story: when it 'dies' it thereby reaches an 'ending'. Hyperfictions always end because readings always end—either with a sense of satisfying closure, or from sheer fatigue.

On the other hand, we are not entirely naive about unresolved texts. Print and cinematic narratives provide instances of multiple closure and also a combination of closure linked to new beginnings. Charles Dickens and other nineteenth-century writers whose novels were serialised in periodicals mastered the art of partial closure in each episode. Furthermore, sequences of novels like Durrell's *Alexandria Quartet* 'suggest that writers of fiction have long encountered problems very similar to those faced by writers of hypertext fiction and have developed an array of formal and thematic solutions to them' (Landow, 1992a:112). The fact that twentieth-century writers and film-makers frequently offer their audiences little in the way of closure indicates that as readers and writers we have long learned to live and read more open-endedly than discussions of narrative form may lead us to believe.

Culturally familiar though we are with the absence or denial of closure, we may still find the consequences disturbing. 'At the moment of pushing narratives beyond the confines and conventions of print', suggests Douglas,

> interactive narratives . . . present readers with a barrage of new and potentially bewildering questions and tasks which promise to redefine our concept of the reader's role. Only further research can answer what do readers do when confronted with narratives without endings or with multiple contradictory endings, how narratives can seem to build sequences from gaps, and how readers traverse intricate networks, suspended in virtual three-dimensional space. (1992:19)

Pedagogical Potential

Rethinking teaching, learning and the curriculum

Writing concerned with the educational promise of hypertext often has an ebullient flavour: there is a lot of 'enthusiastic speculation (and not a little visionary zeal)' (DiPardo and DiPardo, 1990:7). When Lanham (1993:107) argues that 'digitisation of the arts radically democratises them', he places himself firmly in a long tradition of technological utopianism that has made similar claims for the railway, the telephone and the car. Although not always an advocate, Tuman (1992c:4–5) perceives electronic technologies in general and hypertext in particular as 'the source of a possible cultural reorientation as profound in its implications (for literacy education and practically everything else) as the industrial revolution of the last two centuries'. Landow (1992a:160) is convinced 'that even the comparatively limited systems and bodies of literary materials thus far available demonstrate that hypertext and hypermedia have enormous potential to improve teaching and learning'. Passion and enthusiasm aside, Lanham, Tuman, Landow and others focus on important aspects of teaching, learning and the curriculum that may be affected by the use of hypertext. These include the promotion of more independent and active learning, changes to teaching and curriculum practices, and challenges to our assumptions about literacy and literary education. A critical response to the most extravagant claims about the educational promise of hypertext—which smack of technological determinism and utopian myopia—is reserved for the concluding section of this book.

Hypertext is both a teaching and a learning tool. By transferring to students much of the responsibility for accessing, sequencing and deriving meaning from information, hypertext provides an environment in which exploratory or discovery learning may flourish. Hypertext users participate actively when locating information: students become reader-authors, either by choosing individual paths through linked information, or by adding texts and links to the docuverse. Hypertext systems seem to facilitate an implicit, incidental and contextual kind of learning, which is regarded widely as more enduring and transferable than when students are taught directly and explicitly (Vygotsky, 1962; Freedman, 1993).

A hypertext classroom changes the role of the teacher in so far as some of the power and authority is transferred to the students. The teacher becomes something like Bruner's (1986) coach, 'more an older, more experienced partner in a collaboration than an authenticated leader' (Landow, 1992a:123). Students become correspondingly more independent as active shapers of the knowledge they acquire. Using hypertext, teachers are encouraged to present themselves 'in polylogic rather than monologic roles' (McDaid, 1991:218). The writing classroom becomes more like 'a transactional space' (ibid.).

Hypertext is equally appropriate for both individual and collaborative work. An Australian history class might take on an interdisciplinary project about the myths surrounding Gallipoli; a literature class might produce a web about Australian literary history in the 1970s, focusing on drama as a literary genre in the ascendancy. The class would be divided into groups, each responsible for a particular subtopic, such as selected plays by Jack Hibberd and John Romeril. Individual students might be assigned still smaller components, such as the influence of theatre companies like the Australian Performing Group and La Mama. The class as a whole would decide the scope of the topic and how to subdivide it; each group would have to compose and link its own nodes; the whole class would have to meet to map out the hypertext. Individual students would be assessed on their contributions and the connections they made; they might also be graded according to the importance of their group's contribution to the larger hypertext. By adding personally to such a system, every student assumes some responsibility for materials that anyone can use. Together they establish a community of learning, and come to understand that a large part of any investigation rests on the work of others.

The use of hypertext offers opportunities to teach the arts in new ways. For teachers and academics, 'a hypermedia corpus of multidisciplinary materials provides a far more efficient means of developing, preserving, and obtaining access to course materials than has existed before' (Landow, 1992a:123). In addition, hypertext provides a more effective mechanism for storing materials used in previous classes and produced by former students, because it requires less effort to select and to reorganise them. Hypertext also encourages teachers to integrate all the courses in which they are involved. Furthermore, the distinction between 'criticism' and 'creation' becomes, in a digital world, 'a dynamic oscillation: you simply cannot be a critic without being in turn a creator. This oscillation prompts a new type of teaching in which intuitive skills and conceptual reasoning can reinforce one another directly' (Lanham, 1993:107).

Because hypertext easily accommodates interdisciplinary approaches to literary studies, teachers can use it to develop and to extend their students' ability to think critically and to make connections between discrete bodies of information. The electronic facility to make such connections speeds up the processes of skilled reading and creative thinking. The instantaneousness of hypertextual links also permits and encourages sophisticated forms of analysis. Hypertext enables students to assimilate large bodies of information while simultaneously developing those analytic habits they will need in order to think critically about it. Because a hypertext system enables students to integrate materials assembled for one course with those obtained for another, 'the disciplinary boundaries that currently govern academic study of the arts dissolve' (Lanham, 1989:275).

The needs of students with differing abilities can also be met effectively by means of hypertext. Seeing that educational materials do not have to be pitched at a particular standard of ability, verbal and non-verbal materials of varying levels of difficulty can be interwoven and self-paced learning encouraged. Furthermore, students who are less comfortable with the printed book than with alternative information media respond well to hypertext. The technology provides the means of redefining the textbook by renegotiating 'the traditional ratio of alphabetic to iconographic information' (Lanham, 1993:106). Hypertext also provides an alternative means of participation for students who may be inhibited or shy, enabling them to 'contribute' to class discussions by adding to or commenting on an element of the hypertext.

Other students may also be provided for in a hypertext classroom. For distance-education students, hypertext combines the reader's control with the virtual presence of a large number of authors, and thus 'frees the student from the need to be in the physical presence of the teacher' (Landow, 1994:12). Potentially it provides an efficient means for students located anywhere to benefit from materials created at any participating institution. Hypertext can invent 'new kinds of electronic communities capable of nurturing intellectual and other concerns' (ibid.). It also offers an environment that encourages both the individualist and the autodidact.

Hypertext provides opportunities to break the convention of segmenting school and university schedules into units of teaching time. If all the requisite materials are available in a hypertext web, a course need no longer proceed in a linear fashion: in their very first week, students can make links with materials and ideas that normally they would not encounter until the last. Thus hypertext 'frees learners from constraints of scheduling without necessarily destroying the structure and coherence of a course' (Landow, 1992a:132). Projects such as Landow's Intermedia program at Brown University 'stand at odds with our heretofore unquestioned assumption that the curriculum ought to be linear if it possibly could' (Lanham, 1993:133).

In a hypertext classroom, assignments and methods of evaluation need to be considered carefully. 'Whether it is true or not that readers retain less of the information they encounter while reading text on a screen than while reading a printed page, electronically linked text and printed text have different advantages' (Landow, 1992a:133). Ideally, exams and tests should measure the results of a student's use of 'interconnectivity', which is hypertext's 'greatest educational strength as well as its most characteristic feature' (ibid.:134). Exams should not aim simply to test data acquisition. Rethinking assessment in the context of hypertext forces educators to rethink the goals and methods of education.

Teaching writing

Common claims made on behalf of hypertext in the teaching of writing 'include the possibility for promoting associative thinking . . . collaborative learning . . . synthesis in writing from sources . . . distributing traditional authority in texts and classrooms . . . and facilitating deconstructive reading and writing' (Johnson-Eilola, 1992:96). It is difficult, however, to produce research data in support of these claims. Research on hypertext runs into much the same problems as research on word processing: different programs may elicit different types of responses from users (Snyder, 1993). Although Storyspace, a program 'particularly well suited for composition use' (Johnson-Eilola, 1992:96), is becoming more widely used in writing classrooms, the hypertext software to be found in the academy and schools has been designed primarily for the creation of technical and functional documents, and not for educational purposes. At present information about the connections between hypertext and the development of writing remains largely dependent on anecdote and prediction.

Lanham (1993:127) makes 'a few oracular speculations' about the nature of a writing lesson in the age of hypertext:

- the essay will not remain the basic unit of writing instruction
- spelling and grammar-checkers (and basic word-processing facilities such as global search-and-replace, visual isolation of single sentences, and repositioning of sentence elements) will replace the spelling lesson

- punctuation will change: electronic text offers a large 'repertoire of performative signs' (ibid.), 'emoticons' made up of certain letters and symbols used to indicate the spirit in which the text is to be read, as on electronic bulletin boards
- writing will be taught as a three-dimensional rather than a two-dimensional art. Teachers can use the dimensions of colour, image and sound—all 'inextricably intertwined in a dynamic and continually shifting mixture' (ibid.:128).
- the notion of 'remedial' teaching will change with the recognition that some people are better at images or sounds than at words. Once 'expressivity' is conceived of as a single spectrum, there will be no need to stigmatise any particular area of it.

Despite his enthusiasm for hypertext, Johnson-Eilola (1992:118) points out that

> moving from linear text to hypertext is not a simple solution (or even a necessary step in a solution) to the complex problems of teaching and learning writing, not to mention the complexities of the social situations in which computers are used.

With hypertext the writing space is both 'richer' (ibid.) and also more complicated, because hypertext writers must pay more attention than linear writers do to small details and to the overall structure of the text. Readers and writers in hypertext need to attend carefully not only to the verbal text enclosed in the separate boxes on the screen but also to where the box occurs, how the boxes are linked to each other, and much more. Consequently, writers must learn not only verbal but also 'visual rhetoric' (ibid.).

There are other difficulties too. Writers using the technology may become more concerned with 'accumulation than contemplation', for instance, and multiplicity may be 'overwhelming instead of enlightening' (Johnson-Eilola, 1992:119). New users of hypertext—accustomed to the conventions of print media—may feel uncomfortable with the demands of the program, and have to unlearn what they have been taught about writing in order to form new conceptions, particularly about the nature of text and the new ways of interacting with it. 'Changing media is not an easy task; many of us are still struggling to master "the multi-layered grammars of the computer" (Selfe, 1989:3); hypertext literacy is yet another layer, often in conflict with more deeply buried grammars' (ibid.:121).

Whether or not hypertext actually enhances the writing abilities of students, educationists in general and teachers of writing in particular have to look toward an information-laden future. Our education system needs to provide students with those analytical and synthetic abilities which will allow them to recognise, understand and adapt to changes in the environment and the demands they pose. Students need

> the kind of knowledge and skill that will enable them to make sense of their worlds, to determine their own interests, both individual and collective, to see through the manipulations of all sorts of texts in all sorts of media, and to express their own views in some appropriate manner. (Scholes, 1985:15–16)

Rather than a functional literacy that 'allows people to function in society more productively (without questioning the basic inadequacies of the system)', students require a critical literacy that 'encourages people to raise questions about, and perhaps change, current social conditions' (Johnson-Eilola, 1991:101). If we agree that education must aim at teaching people to gather information from a wide variety of sources, and to integrate what they have gathered into a coherent whole so that it becomes knowledge, then hypertext is a useful medium in which to achieve this aim. Both functional and critical literacy can be taught with hypertext. How the technology is to be used will be determined largely by teachers.

Teaching literary studies

The use of hypertext in literary studies is not as yet widespread. Some advocates argue, however, that its impact cannot be ignored when it is integrated into a literature program. In their view, hypertext far exceeds the contextualisation, formal recognition and annotation that have long been the backbone of literary education. For Tuman, hypertext represents both a radically new way of thinking about text and (with the integration of graphics) of organising knowledge itself. Just as the word processor is more than a turbo-charged typewriter, Shakespeare on CD-ROM is more than an efficient vehicle for traditional literary scholarship. Such technologies represent the

> first step toward the creation of a vast hypertext that will subsume all English literature in a single database, dissolving, as it were, the traditional boundaries not just between individual works but between authors and ages as well and altering our existing notions of literature and the act of reading itself. (Tuman, 1992c:5)

In fact, literature courses need not be limited to 'all English literature', since hypertext enables the teaching of literature to become interdisciplinary: all texts—literary or otherwise—are interconnected. Accounts of literary courses that use hypertext technology in creative ways are now beginning to appear. Hypertext enables Landow (1992b) and his students to make complex connections not just between canonical and non-canonical literature, but across the boundaries of such different disciplines as art, music, architecture and history. The technology effortlessly accommodates new work within the total context. By encouraging student participation, team teaching, interdisciplinary approaches and collaborative work, hypertext inevitably redefines the education process, and particularly the development of teaching materials. Being inevitably open-ended and incomplete, hypertext corpora resist closure. Because they can never offer the final word on anything, they avoid appearing authoritative. In such courses—literature, like hypertext—remains an open, changing and expanding system of relationships.

Havholm and Stewart describe how hypertext can be used effectively to explore the ways in which 'ideas and information connect' (1990:8). Aiming to help students 'learn to detect general theoretical assumptions underlying interpretative and critical practice' (ibid.:9), they taught a course focused on intertextuality in the context of seventeenth-century English literature and culture. Using Owl International's Guide, students were able to insert some texts or pictures 'under' others and thus model the three-dimensional implications of intertextuality. At least from the instructors' viewpoint, the course was successful: 'students immersed themselves deeply in a time and culture as they attempted to account for the formation of even individual lines and words of a text' (ibid.:10). The creation of hypertext models also forced them 'to confront the relationships between and among texts' (ibid.). Such exercises stimulated two kinds of active learning: first, instead of merely talking about the application of a theory, students were able to model it; and second, because students were attempting to model the workings of the theory accurately, they were able to infer for themselves some of the consequences of the theory. Havholm and Stewart argue that the operations of literary theory (and of many other theories in the humanities and social sciences) can be simulated by means of hypertext. Moreover, they claim that such modelling requires neither a team of specialists in literary theory nor a group of expert programmers.

Moulthrop and Kaplan (1991, 1994) describe two courses in which they have used hypertext. In the first, they found that the writing of interactive fiction, by raising the possibility of alternative constructions, heightened the sensitivity of their students to narrative features such as point of view, the authority of the narrator and causal sequence. Interactive fiction also seems to help integrate an enriched experience of literature with the practice of writing as a social activity, and enables students to become not merely more perceptive interpreters of fiction, but also creators of it. Interactive hyperfiction, they argue, has considerable potential for those who teach 'writing through literature, or literature through writing' (Kaplan and Moulthrop, 1991:21).

In the second course, an introduction to composition and literature, Moulthrop and Kaplan (1994:225) explored how 'constructive' hypertext could be used 'to engage students in a more open-ended, less constrained encounter with literature'. Kaplan's section used Borges's 'The Garden of Forking Paths', Moulthrop's hypertext version of the story, Joyce's Afternoon and McDaid's Uncle Buddy's Phantom Funhouse (1993). In all three cases, students were faced with narratives that defy closure and encourage highly participatory interactions. The three hypertexts

were also potentially constructive in encouraging revisionary
writing as well as exploratory reading. Both instructors anticipated
resistance to their 'technocritical agenda' (Moulthrop and Kaplan,
1994:228), although not in the form it took when 'a student
undertook a double reversal—to subvert the very subversiveness
of hypertext' (ibid.). Attempting to resist Moulthrop's Forking
Paths 'by objectifying it, [and] establishing an aloofness from its
gregarious metafictional game', he aimed to be 'sober and
reflective, not fictively playful' (ibid.:232). Such 'reader-writerly
resistance' (ibid.:234) could not succeed, according to Moulthrop
and Kaplan, because in hypertext 'there is no way to resist
multiplicity by imposing a univocal and definitive discourse'
(ibid.:235). However many distinctions and hierarchies we try to
introduce, hypertext as a writing system will 'subvert these
constructs' (ibid.:235) by merging everything into the network of
discourses. The imposition of ordering principles on a hypertext
system can never be more than contingent, and susceptible to
reinterpretation or circumvention by subsequent writer-readers.

The introduction of hypertext into literature classrooms raises
a number of pedagogical issues. To treat Afternoon as a literary
text, for instance, involves redefining what is meant by the terms
'literature' and 'text'. Such a reconceptualisation entails not only
those fundamental alterations in the roles of author, reader and
text discussed in earlier chapters, but also changes in the role of
the teacher and in the activities of teaching and learning about
literature. Hypertext foregrounds the interaction between social
concerns and technological developments, and much more so
than in literature classrooms dependent wholly on print-based
materials. But it also raises a problem associated with the use of
large hypertext systems like those created at Brown University,
where user-readers can move freely from text to text, comparing,
cross-referencing and posting responses. Although they overcome
the limitations of a single text, such large bodies of information
may discourage readers from seeking material not available in
electronic form. By making the mistake of assuming that knowledge is merely information retrievable from a computer, they
may overlook gaps and inaccuracies in the information thus
available.

Emergent academic genres

When used to produce and extend familiar forms of discourse, hypertext offers unexpected permutations and hybrid outcomes. But it may also result in new kinds of discourse—dynamic forms of writing that exploit the technology's capacity for making connections in multiple, three-dimensional, complex patterns. Theorists are researching the ways in which hypertext can be used not only to enhance students' writing but also to articulate and to promote new academic genres.

Hypertext promises to assist in 'the cognitively complex task of synthesis writing, or combining information from multiple sources' (Palumbo and Prater, 1993:59). 'Synthesis' requires writers to select from the available content and search for relationships 'that can subsume all or large chunks of information and turn it into a summarised whole' (ibid.:60). The process of selecting, organising and connecting subject matter from a variety of sources teaches students how to construct knowledge from information. As a non-linear, interactive and dynamic medium, hypertext facilitates such processes. The technological features that encourage 'synthesis' include the non-linearity of the knowledge base and its potential for establishing unique, logically related, conceptual webs.

Perhaps even more significantly, hypertext may be changing the form of the academic essay in the humanities and social sciences. As used there for assessment purposes, an essay is 'a work usually defined in terms of its structure or wholeness (its introduction, body, and conclusion all supporting the development of a central thesis) and its content (that its thesis is original or somehow insightful)' (Tuman, 1992a:4). Hypertext makes possible a new kind of essay in which linear argument is replaced by multiple explanations. Argumentation in print involves guiding one's readers through a body of information towards the univocal solution of a problem. Hypertext, by contrast, provides a set of possibilities through which many different arguments or lines might be traced or arranged. Readers of a print-based argument must determine whether or not they are persuaded by the author's point of view; readers of a hypertext must take action at an earlier stage by assuming responsibility themselves for the lines traced through the web. Although these new kinds of texts have caused some concern, they are not necessarily in conflict with the judgements and discriminations we associate with critical thinking and literacy. They represent rather an extension and elaboration of them. Hypertext may well oblige us to redefine essay writing in a way that reveals not just the difficulty but perhaps even the irrelevance of producing a continuous and systematic argument.

Perhaps essays have been at the centre of literacy and literary studies this century mainly because 'as miniature books they reflect the commitment of print culture to the task of generating and comprehending such focused texts' (Tuman, 1992a:4). The central question is not whether books or essays will be entirely replaced by electronic forms, but whether or not they 'will become marginal to the central project of literacy education, much as the formerly dominant practices of memorising texts or copying manuscripts have in the last 200 years become menial tasks that students still routinely perform, albeit usually only in support of other, more central concerns' (ibid.:5).

In speculating about what forms the traditional academic essay may take, we cannot ignore the implications for writing of the incorporation into hypertext webs of graphics and other multimedia dimensions. Tuman asks:

> Is writing going to continue to be what it largely has been, at least for the last century—the linear arrangement of pure text— or is it going to become something else perhaps, text integrated with graphics, either printed as complex documents (with headlines and other graphical cues serving rhetorical purposes) or displayed as a nonlinear series of artfully designed screens? In other words, are we to consider writing mainly in terms of the internal structure of ideas . . . or are we to expand our notion of writing to include pictures, page design and eventually screen presentation . . . ? (1992c:265)

The paradigmatic text of this new age may be the 'polyvocal, interactive, volatile' computer file (Lanham, 1990:xv), a form of expression that has already become 'the primary site of writing in the workplace . . . and . . . may eventually dominate the schools as well' (Tuman, 1992a:4).

According to Ulmer (1992:160), what has emerged from the meeting between print and electronic cultures is a new discursive form: 'collage'. Narrative, exposition and pattern are the dominant ways of organising the release of information in all media. These three modes are not necessarily mutually exclusive; on the contrary, all are present in any work, although one will tend to dominate. Narrative dominates in story, exposition in documentary, and pattern in collage. Ulmer explains:

> One way to construe what is happening as we pass from print apparatus to electronic apparatus would be to say that the dominant forms for organising information in print have been narrative and exposition . . . with pattern dominating only in the arts, at the bottom of the hierarchy of knowledge in the relations among science, social science, and the humanities. The dominant form organising the release of information in the new apparatus, however, is pattern, whose essential form is collage . . . Story and document are still operating in collage, but they are subordinated to and manipulated by the operations of pattern, which transform their signifying effects. (ibid.:160–1)

We should be teaching collage, he continues, by using its own logic and form. Whereas 'narrative . . . reasons abductively, [and] exposition reasons inductively and deductively, . . . pattern reasons conductively'. Conduction has been worked out most thoroughly within the institution of psychoanalysis in terms of dream-work.

> A typical assignment, then, for teaching writing and thinking as pattern primarily, rather than as story or argument primarily, includes an introduction to conductive (dream-work) logic (displacement, condensation, dramatisation, secondary elaboration) as well as to the representational devices of collage (citation or appropriation, juxtaposition, fragmentation). (ibid.: 161)

As hypertext becomes more widely used in the academy, Ulmer's collage may well assume the status of an accepted discourse and perhaps even rival the essay as the pre-eminent form of written communication and assessment.

Articulating a rhetoric of hypertext

Changes to reading and writing practices effected by the use of hypertext oblige us to find ways of talking about the activities it fosters, such as interactive reading, writing and co-authorship. How do we describe a 'document' that has multiple points of entry, multiple exit points and multiple pathways between those points? The development of a multi-linear, interactive hypertext, in which both readers and writers themselves make and remake meaning, requires a new rhetoric: a silicon rhetoric.

The first rhetorics of hypertext 'describe "ordered" and "autonomously meaningful" discursive objects' (Moulthrop, 1991b:151). The purpose of a well-designed hypertext is to provide a supportive environment in which, with a minimal amount of effort, users can find and annotate information, create their own paths through the material and construct webs of information. It was Landow (1991) who first discussed the need for a rhetoric of 'departures' and 'arrivals' that would describe the experience of leaving one hypertextual node and relocating oneself in another.

Carlson (1990) and Shirk (1991) discuss different ways in which hypertext authors and designers can help users to adjust to the new medium. Carlson argues that because we have developed a number of tactics for quickly locating and extracting meaning from linear text any electronic information delivery system 'must be able to duplicate this functionality' (1990:129). In a later article, Carlson advocates advanced forms of graphical browsers as 'essential in overcoming the cognitive load of navigation' (1991:87). Such meta-views of text would help users to see the logical patterns in a body of information and thus assimilate meaning more easily. Shirk believes that creators of hypertext need to know the 'principles of good screen format and design, techniques for creating effective links among nodes . . . of information, and at least the rudiments of computer-user interface design guidelines' (1991:181). In Shirk's view, hypertext authors should keep two considerations in mind when thinking about the presentation of information: first, the need to reduce material 'to a series of discrete units . . . that are presented via a single screen, rather than through a preplanned sequential organisation of pages' (ibid.); secondly, the need 'to develop underlying structures—mental modes or metaphors—for their information' (ibid.:182).

However, Moulthrop warns that we should be wary of attempts to establish absolute principles for linking, navigation and the collaborative creation of hypertext. 'When definitive rhetorics are applied to hypertext they will no doubt operate as they have to date, steering a radically divergent technology back toward the familiar categories and territories of print culture' (1991b:158). Hypertext developers face a dilemma: as users we require rhetorical conventions that limit the complexity of hypertext documents, but such conventions may curtail many of the features that make hypertext so interesting, namely implicit and associational links. The choice lies somewhere between a 'romantic' view of hypertext as a totally unconstrained and freely associative discourse, and an instrumental approach which makes hypertext look not that much different from print. A possible solution, suggests Moulthrop, is to develop 'a *dynamic* rhetoric of hypertext' with 'a strong interdisciplinary character, combining strategies from semiotics, political theory, cognitive science, and phenomenology, among other fields' (ibid.).

Changing approaches to scholarship

Scholars working in the humanities and social sciences have been slow to consider the potential of hypertext. Yet we are rapidly moving into '"a multi-tiered" information universe' (Moulthrop, 1991c:254), where print coexists with electronic forms of text. It will become increasingly difficult for academics to limit their activities to the medium of print, especially when increasing numbers of publishers of books and journals begin to adopt the new technologies. Indeed,

> the combination of electronic textuality and high-speed, broadband networks, which enormously increases the rate of transferring and hence sharing of documents, promises ultimately to reconfigure all aspects of scholarly communication, particularly those involving education and publication. (Landow, 1994:12).

In the academy, hypertext promises 'not simply to streamline our access to writing, but to transform the way we produce and organise bodies of text' (Moulthrop, 1989:19). It 'automates and simplifies the reader's task in moving through a complex, nonlinear document' and, by eliminating 'the distractions of page turning and volume hunting', it allows access to information 'in fractions of a second rather than fractions of an hour or day' (ibid.). This economy of effort 'could substantially alter the pace and scope of intellectual exchange (and not necessarily for the better)' (ibid.). Indeed, as more academics begin to use the World Wide Web—the electronic hypertextual space which accommodates these new scholarly practices—the main problem is going to be not how to acquire texts but how to discern those we value 'from the deafening babble of global electronic traffic' (Delany and Landow, 1993:15).

Hypertext obliges us to look critically at those material properties and conditions which currently characterise institutions of learning. What is to be the future of books, teaching spaces and libraries? What will happen to teachers when they no longer have to front up physically to their students? What is going to happen to the concept of intellectual property, which is very much the product of a print culture, and rests on questionable assumptions about 'originality' and 'textual authority'? It is now clear that 'electronic text and copyright law are steering a collision course at almost every point' (Lanham, 1993:134). All these issues are tremendously important when considering the future of the academy in a computer-dominated culture.

Resistance to the new technologies

The historical record shows that many teachers in the humanities and social sciences have resisted the influence of technology on educational practices. Indeed, many teachers who work in environments that have computer facilities remain reluctant to enter the electronic textual space. They are wary of the use of the technology in education, despite the fact that we face a future dominated by computer culture.

Yet, if the influence of electronic text is to be as pervasive as many are predicting, then all those involved in education need to think about its consequences for teaching and learning in the humanities and social sciences. For as Lanham points out, 'it is hard not to conclude that what we are doing now is not preparing our students for the world they will live in, and the lives they will live out, but training them, instead, to be the "clerks of a forgotten mood"' (1993:136). Addressing teachers of writing and literary studies, Lanham poses the question: 'What business are we really in?' (1989:285). His answer is unequivocal:

> If our business is general literacy, as some of us think, then electronic instructional systems offer the *only* hope for the radically leveraged mass instruction the problems of general literacy pose. If we are in any respect to pretend that 'majoring in English,' or any other literature, and all that it implies, teaches our students how to manipulate words in the world of work, then we must accommodate literary study to the electronic word in which that world will increasingly deal. (ibid.)

In characteristically florid style, Lanham argues that 'electronic technology is full of promising avenues for language instruction of all sorts', and that 'it will be lunacy if we do not construct a sophisticated comparative literature pedagogy upon it' (ibid.:286). Although Lanham is making an important point, his extremism may well alienate those whom he aims to enlist. Such prose is easily dismissed as exaggerated in its claims and evangelical in its exhortations. Moreover, implicit in Lanham's argument is the assumption that if hypertext fails to transform the humanities and social sciences, it will be because of the passivity or ignorance of academics. If his aim is to encourage academics to participate in the new technology, a different approach may be needed.

Assuming a critical perspective

The discussion of hypertext in this book has been conducted mainly from the post-critical position described in the Preface, namely that it is now impossible for educationists to abandon computers—partly because of the enormous economic investment in them, and partly because students see computer knowledge as enhancing their employment opportunities and marketability. Instead of ignoring or demonising computer technology, it is more productive to try to further our understanding of such things as hypertext in order to exploit their educational potential. But it is equally important to bear in mind the reservations expressed by critics who think that the promotion of hypertext is jeopardised by intemperate claims and hype. My positioning of this section at the end of the last chapter is, of course, deliberate. For while we need to know what hypertext is and what it can do, we have to look critically at claims that this innovative technology will radically transform education. Just how excessive the claims of enthusiasts can be is revealed in the satirical title of a conference keynote address: 'Hypertext—Does It Reduce Cholesterol, Too?' (Meyrowitz, 1991).

Central to this book are the affinities between hypertext and the postmodern text as defined by Barthes, Derrida and Foucault: Landow, Bolter, Joyce and others have all drawn parallels between definitions of the postmodern text in contemporary literary theory and the physical characteristics of hypertext. Many of those who enlist postmodern theory point out that key terms used by Barthes and Derrida appear to describe the material properties of hypertext so well that they 'cry out for hypertextuality' (Landow, 1992a:8). For instance, Barthes (1979) describes the text as a network of references to and reflections of other works, and Derrida defines it as 'a differential network, a fabric of traces, referring endlessly to something other than itself, to other differential traces' (1979:84). Hypertext is thus 'celebrated' (Douglas, 1993:419) as a physical instantiation of the concept of the decentred text as put forward by Barthes (1993) and Derrida (1976b), according to which meaning is distributed within a text through a dense network of associations with other texts that lie outside its own physical boundaries. Writers on hypertext often quote Barthes (1979, 1993) on how the role of the reader becomes central after the 'death' of the author 'as the device that controls the signifying potential of the Text' (Douglas, 1993:421).

This book argues on behalf of the educational value of such interconnections: because hypertext *embodies* postmodern theories of the text, it makes it easier to understand them. Other critics, however, are sceptical. Although herself an exponent of the 'embodiment' view, Douglas (1993:423) observes (with more than a little self-parody) that the use of postmodern literary theory by apologists for hypertext is honorific, in that it serves 'to make preliminary descriptions of the new environment appear more substantial by anchoring them to a large and highly esteemed body of critical writing'. By claiming that Barthes and Derrida 'uncannily anticipated' (ibid.:423) electronic text in their definitions of the postmodern text, theorists are able to situate hypertext in a discursive continuum from the spoken word right through to hypertext itself.

Grusin (1994:475), on the other hand, attacks the 'embodiment' argument as a pointless exercise, in so far as literary theories like post-structuralism, postmodernism and deconstruction do 'not need to be instantiated or embodied in new electronic technologies'. The force of the Derridean critique is 'to demonstrate the way in which thought and speech are always already forms of writing': for Derrida, writing is 'always a technology and already electronic'. Grusin believes that Barthes' distinctions between 'work' and 'text' or between 'readerly' and 'writerly' texts—both of which are 'habitually cited as theoretical anticipations of the technology of electronic writing' (ibid.)—have been similarly misread. For Barthes as for Derrida, 'the 'writerly' text is always already immaterial, allusive and intertextual—even in print'. As Grusin goes on to argue,

> this is not to deny that in electronic writing the 'work' has taken a different form, one that seems more closely to resemble the Barthesian 'text'. But in describing hypertext or electronic writing as embodying the assumptions of 'Barthesian post-structuralism or Derridean deconstruction, electronic enthusiasts run the risk of fetishising the 'work', of mistaking the 'work' for the 'text', the physical manifestation (electronic technologies) for the linguistic or discursive text . . . The force of the decon-structive and poststructuralist critiques is to illustrate the way in which this destabilisation is true of all writing. To think otherwise is not to instantiate or embody these critiques but to mistake or ignore them. (ibid.)

Yet although 'critical theorists describing decentered, polyvocal, and nonlinear forms of writing were specifically arguing about shifts in the status of discourse and the social context surrounding the production and consumption of printed texts' (Douglas, 1993:421), the applicability of such critiques to an understanding of hypertext is not really at issue. If by pointing to these connections we can help students to understand such theories, then it becomes a useful exercise. More problematic, however, is the way in which theorists writing about hypertext use literary theory 'to make predictions about the social impact of hypertext' (ibid.). We need to examine very carefully the ways in which the 'embodiment' argument can be appropriated in the interests of a technological determinism that sees hypertext as inevitably transforming both society and its education systems. We need a critical perspective to understand that 'many of the assessments of the potentially "revolutionary" impact of hypertext on society are overly simplistic' (ibid.:417).

Both Landow and Lanham enthusiastically endorse the consequences of hypertext for education in the humanities and social sciences. Landow claims that hypertext challenges conventional assumptions about teachers, learners and institutions, and reconfigures everything. But he is not optimistic of this happening in the immediate future, given our technological conservatism and a lack of concern about pedagogy. He is also careful to qualify the importance of links between hypertext and literary theory, preferring to view the text as a site on which the accuracy of these theoretical concepts can be tested. By contrast, Lanham predicts that hypertext will result in a far-reaching social revolution, in which education and the arts will be radically democratised, and minority groups enfranchised. The problem with such claims is their assumption 'that society will be shaped largely by the widespread use of hypertext technology, not that the developing technology will be moulded by social interests' (Douglas, 1993: 421). Lanham's perspective is technocentric in that it

> permits theorists to make judgements about the type of interactions readers and writers will have with hypertext as well as predictions about its effects on the institutions of publishing, education and government—all based on the physical characteristics of hypertext as an environment for discourse. (ibid.:423)

Incipient technological determinism is discernible in many evocations of the educational possibilities of hypertext and manifests itself in 'propositional statements that ascribe agency to technology itself' (Grusin, 1994:470). It is present also in claims that the electronic media and network technologies could have deleterious effects on human- consciousness and culture by splintering and isolating both groups and individuals (Bigum and Green, 1993). Technological determinists who predict the social consequences of hypertext tend to rely vulnerably on either a utopian or a dystopian view of the future. But because hypertext can be used for all sorts of purposes, it can both 'liberate' and 'constrain' educational and social practices.

No technology can guarantee any particular behaviour simply by its 'nature'. A hypertext classroom can be used either to support new theories of reading and writing or to promote traditional approaches to the study of texts. Teachers who are neither trained in nor sympathetic towards hypertext pedagogy will either ignore or subvert the potential of that technology. The use and effect of a technology is closely tied to the social context in which it appears. Hypertext will succeed or fail not by its own agency but as a result of how it is used by the people and institutions that take it up. Every evolving technology is socially constructed in the sense that it is shaped by the interests and assumptions of particular social groups. It will be interesting to see how teachers in the humanities and social sciences respond to the possibilities afforded by hypertext. What we have at present is an intriguing technology whose operations compel us to radically revise our print-derived notions of reading, writing, text, language and closure. My problem now is how to conclude an analysis of a technology that itself defies closure. But since I am writing to a contracted word-length in a bound volume that not only invites but requires closure, my 'difficulty' is no more than a construction. Perhaps it is enough to say that hypertext marks the beginning of a new kind of communication. The technology offers an opportunity for teachers and students to produce, circulate and receive texts in an unparalleled and exciting confluence of literature, writing and technology.

Bibliography

Amato, J. (1991) 'Review of Writing Space: The Computer, Hypertext, and the History of Writing', *Computers and Composition* 9, 1, pp. 11–17

Aristotle (1959) *The Poetics*, trans. L. J. Potts, *Aristotle on the Art of Fiction: An English Translation of Aristotle's* Poetics *with an Introductory Essay and Explanatory Notes*, Cambridge, United Kingdom: Cambridge University Press

Aronowitz, S. (1992) 'Looking Out: The Impact of Computers on the Lives of Professionals' in Tuman (ed.) *Literacy Online*, pp. 119–37

Bakhtin, M. M. (1981) *The Dialogic Imagination*, Austin, Texas: Texas University Press [Rus 1975]

—— (1984) *Problems of Dostoevsky's Poetics*, ed. and trans. C. Emerson, Minneapolis: University of Minnesota Press [Rus 1963]

Balestri, D. P. (1988) 'Softcopy and Hard: Word Processing and Writing Process', *Academic Computing* 2, 5, pp. 14–17, 41–5

Barth, J. (1967) 'The Literature of Exhaustion', *Atlantic* 220, 2, pp. 29–34

Barthes, R. (1974) *S/Z*, trans. R. Miller, Oxford: Blackwell [Fr 1973]

—— (1975) *The Pleasure of the Text*, trans. Richard Miller, New York: Hill and Wang [Fr 1973]

—— (1979) 'From Work to Text', trans. J. Harari, in Harari (ed.) *Textual Strategies*, pp. 73–81 [Fr 1971]

—— (1986) *The Rustle of Language*, trans. R. Howard, New York: Hill and Wang [Fr 1984]

—— (1993) 'The Death of the Author', trans. S. Heath, in Rice and Waugh (eds) *Modern Literary Theory: A Reader*, 2nd edn, pp. 114–21 [Fr 1968]

Berk, E. and J. Devlin (1991) (eds) *Hypertext/Hypermedia Handbook*, New York: Intertext Publications McGraw-Hill

Bernstein, M. (1991) 'The Navigation Problem Reconsidered' in Berk and Devlin (eds) *Hypertext/Hypermedia Handbook*, pp. 285–97

Bigum, C. and B. Green (1993) 'Technologising Literacy: or, Interrupting the Dream of Reason' in A. Luke and P. Gilbert (eds) *Literacy in Contexts: Australian Perspectives and Issues*, Sydney: Allen and Unwin, pp. 4–28

Bolter, J. D. (1991) *Writing Space: The Computer, Hypertext, and the History of Writing*, Hillsdale, New Jersey: Lawrence Erlbaum Associates

—— (1992a) 'Literature in the Electronic Writing Space' in Tuman (ed.) *Literacy Online*, pp. 19–42

—— (1992b) 'Discussion' in Tuman (ed.) *Literacy Online*, pp. 59–63

—— (1993) 'Alone and Together in the Electronic Bazaar', *Computers and Composition* 10, 2, pp. 5–18

Bolter, J. D., M. Joyce and J. B. Smith (1990) *Storyspace: Hypertext Writing Environment for the Macintosh*, computer software, Cambridge, Massachusetts: Eastgate Systems

Borges, J. L. (1970a) *Labyrinths*, D. A. Yates and J. E. Irby (eds) Harmondsworth, Middlesex: Penguin Books

—— (1970b) 'The Library of Babel' in *Labyrinths*, pp. 78–85

—— (1970c) 'The Garden of Forking Paths' in *Labyrinths*, pp. 44–54

—— (1977) *The Book of Sand*, trans. Norman Thomas di Giovanni, New York: Dutton [Sp 1975]

Bruner, J. (1986) *Actual Minds, Possible Worlds*, Cambridge, Massachusetts: Harvard University Press

Bush, V. (1945) 'As We May Think' in Nyce and Kahn (eds) *From Memex to Hypertext*, pp. 85–107

Calder, J. (1982) 'Introduction' in J. Calder (ed.) *A William Burroughs Reader*, London: Picador, pp. 7–24

Carlson, P. A. (1990) 'The Rhetoric of Hypertext', *Hypermedia* 2, 2, pp. 109–31

—— (1991) 'New Metaphors for Electronic Text', *Impact Assessment Bulletin* 9, 1–2, pp. 73–91

Catano, J. (1985) 'Computer-based Writing: Navigating the Fluid Text', *College Composition and Communication* 36, pp. 309–16

Coleridge, S. T. (1849) 'General Introduction; or, A Preliminary Treatise on Method' in *Encyclopedia Metropolitana,* ed. E. Smedley, H. J. Rose and J. Rose, vol. 1, London: John Joseph Griffin, pp. 1–28

Collison, R. (1966) *Encyclopedias: Their History Throughout the Ages*, New York: Hafner

Cortazar, J. (1966) *Hopscotch*, trans. Gregory Rabassa, London: Collins Harvill [Sp 1963]

Culler, J. (1981) *The Pursuit of Signs: Semiotics, Literature, Deconstruction*, Ithaca, New York: Cornell University Press

Dede, C. J. and D. B. Palumbo (1991) 'Implications of Hypermedia for Human Thought and Communication', *Impact Assessment Bulletin* 9, 1–2, pp. 15–28

Delany, P. and G. P. Landow (1991) (eds) *Hypermedia and Literary Studies*, Cambridge, Massachusetts: MIT Press

—— (1993) 'Managing the Digital Word: The Text in an Age of Electronic Reproduction' in Landow and Delany (eds) *The Digital Word*, pp. 3–28

Deleuze, G. and F. Guattari (1987) *A Thousand Plateaus: Capitalism and Schizophrenia*, trans. B. Massumi, Minneapolis: University of Minnesota Press

Derrida, J. (1973) *Speech and Phenomena*, trans. D. B. Allison, Evanston, Illinois: Northwestern University Press [Fr 1967]

—— (1976a) *Glas*, trans. J. P. Leavey, Jr, and R. Rand, Lincoln, Nebraska: University of Nebraska Press [Fr 1974]

—— (1976b) *Of Grammatology*, trans. G. Chakravorty Spivak, Baltimore, Maryland: Johns Hopkins University Press [Fr 1967]

—— (1977) 'Signature Event Context', *Glyph* 1: Johns Hopkins Textual Studies, Baltimore, Maryland: Johns Hopkins University Press, pp. 172–97

—— (1979) 'Living On' in J. Hulbart (ed.) *Deconstruction and Criticism*, New York: Seabury Press, pp. 75–176

DiPardo, A. and M. DiPardo (1990) 'Towards the Metapersonal Essay: Exploring the Potential of Hypertext in the Composition Class', *Computers and Composition* 7, 3, pp. 7–22

Douglas, J. Y. (1989) 'Wandering Through the Labyrinth: Encountering Interactive Fiction', *Computers and Composition* 6, 3, pp. 93–101

—— (1991) 'Understanding the Act of Reading: the WOE Beginner's Guide to Dissection', *Writing on the Edge* 2, 2, pp. 112–25

—— (1992) 'What Hypertexts Can Do the Print Narratives Cannot', *Reader* 28, pp. 1–22

—— (1993) 'Social Impacts of Computing: The Framing of Hypertext—Revolutionary for Whom?', *Social Science Computer Review* 2, 4, pp. 417–29

Drexler, K. E. (1986) *Engines of Creation*, New York: Anchor Press/Doubleday

Eco, U. (1979) *The Role of the Reader*, Bloomington, Indiana: Indiana University Press

Ellul, J. (1980) *The Technological Society*, New York: Alfred A. Knopf

Encyclopaedia Britannica (1974–87), ed. P. W. Goetz, Chicago: Encyclopaedia Britannica

Encyclopedia Metropolitana (1849) ed. E. Smedley, H. J. Rose and J. Rose, vol. 1, London: John Joseph Griffin

Fish, S. (1980) *Is There a Text in This Class? The Authority of Interpretive Communities*, Cambridge, Massachusetts: Harvard University Press

Foucault, M. (1976) *The Archeology of Knowledge*, trans. A. M. Sheridan Smith, New York: Harper Colophon [Fr 1966]

—— (1979) 'What is an Author?' in Harari (ed.) *Textual Strategies*, pp. 141–60 [Fr 1969]

Freedman, A. (1993) 'Show and Tell? The Role of Explicit Teaching in the Learning of New Genres', *Research in the Teaching of English* 27, 3, pp. 222–51

Grolier Electronic Publishing (1988) *The Electronic Encyclopedia on CD-ROM*, Danbury, Connecticut: Grolier Electronic Publishing

Grusin, R. (1994) 'What is an Electronic Author? Theory and the Technological Fallacy', *Configurations* 3, pp. 469–83

Guyer, C. and M. Petry (1991) Izme Pass, computer disk, *Writing on the Edge* 2, 2

Hager, T. (1993) 'A Rhetoric of Hypertextual *Inventio*', *Writing on the Edge* 4, 2, pp. 103–15

Harari, J. (1979) (ed.) *Textual Strategies*, Ithaca, New York: Cornell University Press

Haraway, D. (1991) 'A Cyborg Manifesto: Science, Technology, and Socialist Feminism in the Late Twentieth Century', *Simians, Cyborgs, and Women: The Reinvention of Nature*, New York: Routledge, pp. 149–81

Harpold, T. (1991) 'The Contingencies of the Hypertext Link', *Writing on the Edge* 2, 2, pp. 126–37

Havholm, P. and L. Stewart (1990) 'Modeling the Operation of Theory', *Academic Computing* 4, 6, pp. 8–12, 46–8

Hawisher, G. E. and C. L. Selfe (1991) (eds) *Evolving Perspectives on Computers and Composition Studies: Questions for the 1990s*, Urbana, Illinois: National Council of Teachers of English

Heim, M. (1987) *Electric Language: A Philosophical Study of Word Processing*, New Haven, Connecticut: Yale University Press

Hirsch, E. D., Jr, (1987) *Cultural Literacy: What Every American Needs to Know*, Boston: Houghton Mifflin

Iser, W. (1980) 'The Reading Process: A Phenomenological Approach' in J. Tompkins (ed.) *Reader-response Criticism: From Formalism to Post-structuralism*, Baltimore, Maryland: Johns Hopkins University Press, pp. 50–69

Johnson-Eilola, J. (1991) '"Trying to See the Garden": Interdisciplinary Perspectives on Hypertext Use in Composition Instruction', *Writing on the Edge* 2, 2, pp. 92–111

—— (1992) 'Structure and Text: Writing Space and Storyspace', *Computers and Composition* 9, 2, pp. 95–129

—— (1993) 'Control and the Cyborg: Writing and Being Written in Hypertext', *Journal of Advanced Composition* 13, 2, pp. 381–99

—— (1994) 'Reading and Writing in Hypertext: Vertigo and Euphoria' in Selfe and Hilligoss (eds) *Literacy and Computers*, pp. 195–219

Joyce, M. (1991a) 'Selfish Interaction or Subversive Texts and the Multiple Novel' in Berk and Devlin (eds) *Hypertext/Hypermedia Handbook*, pp. 79–92

—— (1991b) Afternoon, a Story, computer disk, Cambridge, Massachusetts: Eastgate Press

—— (1991c) WOE, computer disk, *Writing on the Edge* 2, 2

—— (1992) 'A Feel for Prose: Interstitial Links and the Contours of Hypertext', *Writing on the Edge* 4, 1, pp. 83–101

—— (1995a) *Of Two Minds: Hypertext Pedagogy and Poetics*, Ann Arbor: University of Michigan Press

—— (1995b) 'Hypertext and Hypermedia' in *Of Two Minds*, pp. 19–29

—— (1995c) 'Siren Shapes; Exploratory and Constructive Hypertexts' in *Of Two Minds*, pp. 39–59

—— (1995d) 'Introduction: The Comfort of Knowing We Are Not Lost' in *Of Two Minds*, pp. 1–15

Kaplan, N. and S. Moulthrop (1991) 'Something to Imagine: Literature, Composition, and Interactive Fiction', *Computers and Composition* 9, 1, pp. 7–23

Kelly, K. (1994) *Out of Control: The New Biology of Machines*, London: Fourth Estate

Landow, G. P. (1991) 'The Rhetoric of Hypermedia: Some Rules for Authors' in Delany and Landow (eds) *Hypermedia and Literary Studies*, pp. 81–104

—— (1992a) *Hypertext: The Convergence of Contemporary Critical Theory and Technology*, Baltimore, Maryland: Johns Hopkins University Press

—— (1992b) 'Hypertext, Metatext, and the Electronic Canon' in Tuman (ed.) *Literacy Online*, pp. 67–94

—— (1994) 'What's a Critic to Do? Critical Theory in the Age of Hypertext' in G. P. Landow (ed.) *Hyper/Text/Theory*, Baltimore, Maryland: Johns Hopkins University Press, pp. 1–48

Landow, G. P. and P. Delany (1991) 'Hypertext, Hypermedia, and Literary Studies: The State of the Art' in Delany and Landow (eds) *Hypermedia and Literary Studies*, pp. 1–50

—— (1993) (eds) *The Digital Word: Text-based Computing in the Humanities*, Cambridge, Massachusetts: MIT Press

Lanham, R. A. (1989) 'The Electronic Word: Literary Study and the Digital Revolution', *New Literary History* 20, pp. 265–90

—— (1990) 'Foreword' in C. Handa (ed.) *Computers and Community: Teaching Composition in the Twenty-first Century*, Portsmouth, New Hampshire: Boynton Cook, pp. xiii–xv

—— (1993) *The Electronic Word: Democracy, Technology and the Arts*, Chicago: University of Chicago Press

McArthur, T. (1986) *Worlds of Reference: Lexicography, Learning and Language from the Clay Tablet to the Computer*, Cambridge, United Kingdom: Cambridge University Press

McDaid, J. (1991) 'Toward an Ecology of Hypermedia' in Hawisher and Selfe (eds) *Evolving Perspectives on Computers and Composition Studies*, pp. 203–23

—— (1993) Uncle Buddy's Phantom Funhouse, computer disk, Cambridge, Massachusetts: Eastgate Press

McLuhan, M. (1964) *Understanding Media: The Extensions of Man*, London: Routledge and Kegan Paul

Meyrowitz, N. (1991) 'Hypertext—Does It Reduce Cholesterol, Too?' in Nyce and Kahn (eds) *From Memex to Hypertext*, pp. 287–318

Moulthrop, S. (1986) Forking Paths, computer disk, unpublished hypertext

—— (1989) 'In the Zones: Hypertext and the Politics of Interpretation', *Writing on the Edge* 1, 1, pp. 18–27

—— (1991a) 'Toward a Paradigm for Reading Hypertexts: Making Nothing Happen in Hypermedia Fiction' in Berk and Devlin (eds) *Hypertext/Hypermedia Handbook*, pp. 65–78

—— (1991b) 'Polymers, Paranoia, and the Rhetoric of Hypertext', *Writing on the Edge* 2, 2, pp. 150–9

—— (1991c) 'The Politics of Hypertext' in Hawisher and Selfe (eds) *Evolving Perspectives on Computers and Composition Studies*, pp. 253–71

—— (1991d) Victory Garden, computer disk, Cambridge, Massachusetts: Eastgate Press

—— (1992) 'Informand and Rhetoric: A Hypertextual Experiment', *Writing on the Edge* 4, 1, pp. 103–27

Moulthrop, S. and N. Kaplan (1994) 'They Became What They Beheld: The Futility of Resistance in the Space of Electronic Writing' in Selfe and Hilligoss (eds) *Literacy and Computers*, pp. 220–37

Nelson, T. H. (1978) *Computer Lib/Dream Machines*, South Bend, Indiana: The Distributors

—— (1992a) *Literary Machines 93.1*, Sausalito, California: Mindful Press

—— (1992b) 'Opening Hypertext: A Memoir' in Tuman (ed.) *Literacy Online*, pp. 43–57

Nyce, J. M. and P. Kahn (1991) (eds) *From Memex to Hypertext: Vannevar Bush and the Mind's Machine*, San Diego: Academic Press

Ong, W. J. (1982) *Orality and Literacy: The Technologizing of the Word*, London: Methuen

Pagels, H. R. (1989) *The Dreams of Reason: The Computer and the Rise of the Sciences of Complexity*, New York: Bantam

Palumbo, D. B. and D. Prater (1993) 'The Role of Hypermedia in Synthesis Writing', *Computers and Composition* 10, 2, pp. 59–70

Papert, S. (1980) *Mindstorms: Children, Computers, and Powerful Ideas*, New York: Harvester Press

Poster, M. (1990) *The Mode of Information: Poststructuralism and Social Context*, Cambridge, United Kingdom: Polity Press

Rice, P. and P. Waugh (1993) (eds) *Modern Literary Theory: A Reader*, 2nd edn, London: Edward Arnold

Ricoeur, P. (1984) *Time and Narrative*, trans. K. McLaughlin and D. Pellauer, vol. 1, Chicago: University of Chicago Press [Fr 1983]

Said, E. W. (1985) *Beginnings: Intention and Method*, New York: Columbia University Press

Scholes, R. (1985) *Textual Power*, New Haven, Connecticut: Yale University Press

—— (1986) 'Aiming a Canon at the Curriculum', *Salmagundi* 72, pp. 101–17

Selfe, C. L. (1989) 'Redefining Literacy: The Multilayered Grammars of Computers' in Hawisher and Selfe (eds) *Critical Perspectives on Computers and Composition Instruction*, New York: Teachers College Press, pp. 3–15

Selfe, C. L. and S. Hilligoss (1994) (eds) *Literacy and Computers: The Complications of Teaching and Learning with Technology*, New York: Modern Language Association of America

Selfe, C. L. and R. J. Selfe (1994) 'The Politics of the Interface: Power and its Exercise in Electronic Contact Zones', *College Composition and Communication* 45, 4, pp. 480–504

Shirk, H. N. (1991) 'Hypertext and Composition Studies' in Hawisher and Selfe (eds) *Evolving Perspectives on Computers and Composition Studies*, pp. 177–202

Slatin, J. M. (1990) 'Reading Hypertext: Order and Coherence in a New Medium', *College English* 52, 8, pp. 870–83

Smith, B. H. (1968) *Poetic Closure: A Study of How Poems End*, Chicago: University of Chicago Press

Snyder, I. A. (1993) 'Writing with Word Processors: A Research Overview', *Educational Research* 35, 1, pp. 49–68

—— (1994) 'Re-inventing Writing with Computers', *Australian Journal of Language and Literacy* 17, 3, pp. 182–97

Sterne, L. (1987) *The Life and Opinions of Tristram Shandy, Gentleman*, ed. I. Campbell Ross, Oxford: Oxford University Press [1759–67]

Strick, P. (1994) 'Smoking/No Smoking', *Sight and Sound* 9, pp. 46–8

Tuman, M. C. (1992a) *Word Perfect: Literacy in the Computer Age*, London: Falmer Press

—— (1992b) (ed.) *Literacy Online*, Pittsburgh, Pennsylvania: University of Pittsburgh Press, pp. 119–37

—— (1992c) 'First Thoughts' in Tuman (ed.) *Literacy Online*, pp. 3–15

Ulmer, G. L. (1985) *Applied Grammatology: Post(e)-Pedagogy from Jacques Derrida to Joseph Beuys*, Baltimore, Maryland: Johns Hopkins University Press

—— (1989) *Teletheory: Grammatology in the Age of Video*, London: Routledge

—— (1992) 'Grammatology (in the Stacks) of Hypermedia' in Tuman (ed.) *Literacy Online*, pp. 139–64

Vygotsky, L. (1962) *Thought and Language*, trans. E. Hanfmann and G. Vakar, Cambridge, Massachusetts: MIT Press [Rus 1934]

Weizenbaum, J. (1976) *Computer Power and Human Reason: From Judgement to Calculation*, New York: W. H. Freeman

Woolley, B. (1992) *Virtual Worlds: A Journey in Hype and Hyper-reality*, Harmondsworth, Middlesex: Penguin Books

Zuboff, S. (1984) *In the Age of the Smart Machine: The Future of Work and Power*, New York: Basic Books

Index